SpringerBriefs in Molecular Science

W0037314

For further volumes:
http://www.springer.com/series/8898

Silvia A. Brandán

A Structural and Vibrational Investigation into Chromylazide, Acetate, Perchlorate, and Thiocyanate Compounds

 Springer

Silvia A. Brandán
Facultad de Bioquímica,
Instituto de Química Inorgánica,
  Química y Farmacia
Universidad Nacional de Tucumán
Tucumán
Argentina

ISSN 2191-5407   ISSN 2191-5415   (electronic)
ISBN 978-94-007-5753-0   ISBN 978-94-007-5754-7   (eBook)
DOI 10.1007/978-94-007-5754-7
Springer Dordrecht Heidelberg New York London

Library of Congress Control Number: 2012952013

Printed on acid-free paper

Springer is part of Springer Science+Business Media (www.springer.com)

Contents

Chapter 1
Structural and Vibrational Properties of Chromyl Azide

Abstract In this chapter, the structural and vibrational properties of chromyl azide were studied from a theoretical point of view using density functional theory (DFT) methods. The initial geometry was fully optimized at different theory levels and the harmonic wavenumbers were evaluated at the same levels. These results show that for the compound, a stable molecule was theoretically determined in the gas phase. Also, the characteristics and nature of the Cr–O and Cr ← O bonds for the stable structure were studied through the Wiberg's indexes calculated by means of the natural bond orbital (NBO) study. On the other hand, the corresponding topological properties of the electronic charge density were analyzed using Bader's atoms in the Molecules theory (AIM). The results were used to predict the infrared and Raman spectra and the molecular geometry of the compound, for which there are no experimental data. Besides, a complete assignment of all observed bands in the infrared spectrum for the compound was performed combining DFT calculations with Pulay's scaled quantum mechanics force field (SQMFF) methodology. Therefore, these calculations gave us a precise knowledge of the normal modes of vibration taking into account the type of coordination adopted by azide ligands of this compound as monodentate. In this chapter, the scaled force constants and the scaling factors are also reported together with a comparison of the obtained values for similar compounds.

Keywords Chromyl azide · Vibrational spectra · Molecular structure · Force field · DFT calculations

1.1 Introduction

For a long time the physical and chemical properties of inorganic azides have been reported by numerous authors [1–14]. These compounds are extensively used in organic syntheses [2–5]. The electronic structure of lithium, sodium, and lead azides

S. A. Brandán, *A Structural and Vibrational Investigation into Chromylazide, Acetate, Perchlorate, and Thiocyanate Compounds*, SpringerBriefs in Molecular Science, DOI: 10.1007/978-94-007-5754-7_1, © The Author(s) 2013

were analyzed by Younk et al. [6] in order to demonstrate the importance of solid-state effects on the electronic structure and possible behavior of such energetic systems. Recently, a detailed ab initio study on the electronic structure and optical properties of crystalline strontium and barium azides have been performed by Zhu et al. [7] using the DFT method within the generalized gradient approximation (GGA). The infrared spectrum of sodium azide was initially used as an internal standard for quantitative infrared analyses [8], new absorption bands appearing in that spectrum when sodium azide crystals are irradiated with ultraviolet light at 77° K [9]. At present, this compound is important as a preservative on raw milk components [10] and as a potent vasodilator that causes severe hypotension on accidental exposure [11]. The synthesis for chromyl azide, $CrO_2(N_3)_2$ and its ultraviolet visible spectrum in a tetrachloride solution was reported by Krauss et al. [15, 16]. So far, there is no theoretical study concerning either geometry or vibrational spectra for chromyl azide, so a comparative work was performed in this chapter intended to evaluate the best theoretical level and basis set to reproduce the experimental data existing for similar compounds that contain azide and chromyl groups in their structures in order to predict the molecular geometry and vibrational spectra of chromyl azide for which no experimental data exist. For that purpose, the optimized geometries and wavenumbers for the normal modes of vibration of sodium azide [12–14], nitride anion, chromyl chloride [17, 18], and fluoride [19, 20] compounds were calculated. The harmonic force constants given by these calculations were subsequently scaled to reproduce as well as possible the proposed experimental wavenumbers obtained from those similar molecules. Then, the results for similar compounds were compared with those obtained for chromyl azide. In addition, the nature of the different types of bonds and their corresponding topological properties of the electronic charge density were systematic and quantitatively investigated using the NBO analysis [21–23] and the atoms in molecules theory (AIM) [24]. The theoretical electronic spectrum of the substance is also compared with the corresponding reported [15] and, then, a complete assignment of the observed bands is shown.

1.2 Structural Analysis

For chromyl azide, a stable structure of C_2 symmetry was found, as shown in Fig. 1.1. Table 1.1 shows a comparison of the total energies and dipole moment values for chromyl azide with the corresponding values for sodium azide and nitride anion using the Lanl2dz, STO-3G*, 6-31G*, 6-31+G*, 6-311G*, and 6-311+G* basis sets and the B3LYP method. The optimized structure for chromyl azide has C_2 symmetry, while the corresponding structures for sodium azide and nitride anion have C_S and $D_{\infty h}$ symmetries, respectively. For chromyl azide, the highest dipole moment value and the most stable structure are obtained with the B3LYP/6-311G* method, while for sodium azide and nitride anion the most stable structures are obtained with the B3LYP/6-311+G* method. For these two species,

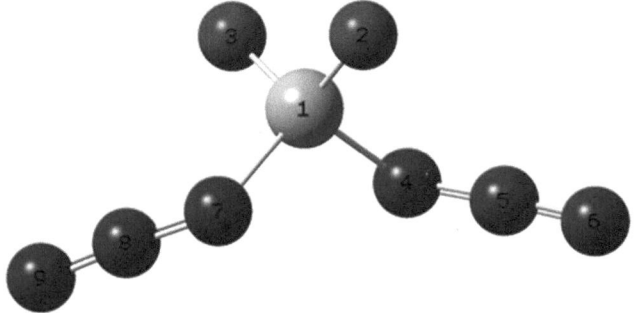

Fig. 1.1 The C_2 molecular structure of chromyl azide considered

the highest dipole moment values are obtained using basis sets with diffusion function, these being the 6-31+G* and 6-311+G* basis sets (Table 1.1).

A comparison of experimental data for chromyl chloride [18] and fluoride [19] compounds, hydrogen [25] and sodium [12, 13] azides, and nitride anion [12, 25] with the calculated geometrical parameters for chromyl azide using a 6-31G* basis set can be seen, respectively, in Table 1.2. According to these results, the experimental geometrical parameters that best reproduce the compound with C_2 structure are those corresponding to chromyl fluoride [19] with C_{2v} structure where the mean difference for bond lengths is 0.085 Å, while it is 3.5° for angles.

On the other hand, the experimental distances corresponding to the HN_3 [25] compound are those that best reproduce the calculated Cr–N–N and N–N bond lengths of the compound where the mean difference for bond lengths is 0.011 Å. While the MNN and NNN angles values that best reproduce the compound are those calculated with B3LYP/6-31G* method for sodium azide where the mean difference is 36.2° for angles. Note that the B3LYP/6-31G* calculations for chromyl azide predict that the O=Cr=O angle is higher than the N–Cr–N bond angle in accordance with the valence-shell electron-pair repulsion (VSEPR) theory [26, 27].

Table 1.1 Total energy (*ET*) and dipole moment (μ) for chromyl and sodium azides and nitride ion using different basis sets

B3LYP method						
Basis set	$CrO_2(N_3)_2$		$Na(N_3)$		N_3^-	
	C_2 symmetry		C_S symmetry		$D_{\infty h}$ symmetry	
	ET (Hartree)	μ (D)	ET (Hartree)	μ (D)	ET (Hartree)	μ (D)
LanL2DZ	−565.08362	1.09	−164.38869	11.58	−164.18138	0.22
STO-3G*	−1502.39646	0.43	−322.35179	7.76	−161.88698	0.11
6-31G*	−1523.29622	0.75	−326.51407	10.72	−164.21316	0.19
6-31 + G*	−1523.31968	0.83	−326.52766	11.48	−164.24484	0.30
6-311G*	−1523.47733	1.20	−326.56731	10.85	−164.26229	0.20
6-311 + G*	−1523.33112	1.05	−326.57502	11.43	−164.28568	0.30

Table 1.2 Comparison of calculated geometrical parameters for chromyl azide compared with the corresponding experimental values for similar compounds

Parameter	C_2 symmetry	C_{2v} symmetry	
	$CrO_2(N_3)_2^a$	$CrO_2Cl_2^b$	$CrO_2F_2^c$
$r(Cr–O)$ (Å)	1.555	1.577	1.572
$r(Cr-X)$ (Å)	1.835	2.122	1.716
RMSD	–	0.203	0.085
$\theta(OCrO)$ (°)	111.8	108.5	107.8
$\theta(OCrX)$ (°)	107.9	108.7	109.3
$\theta(XCrX)$ (°)	109.3	113.2	111.9
RMSD	–	3.6	3.5
Parameter	$CrO_2(N_3)_2^a$	C_S symmetry	
		HN_3^d	NaN_3^e
$r(_M–N–N)$ (Å)	1.231	1.241	1.173,1.201[a]
$r(N–N)$ (Å)	1.140	1.128	1.167,1.160[a]
RMSD	–	0.011	0.049, 0.025
$\theta(NNN)$ (°)	174.4	112.6	180.0[a]
$\theta(MNN)$ (°)	125.5	110.9	176.4[a]
RMSD	–	44.9	36.2

[a] This work B3LYP/6-31G*
[b] Ref [18]
[c] Ref [19]
[d] Ref [25]
[e] Ref [14]

The intermolecular interactions for chromyl azide have been analyzed using Bader's topological analysis of the charge electron density, $\rho(r)$, by using the AIM program [28]. For the characterization of the compound molecular electronic structure, the determination of the $\rho(r)$ in the bond critical points (BCPs) and the values of the Laplacian parameter at these points were determined and compared with those calculated for chromyl chloride and fluoride compounds, hydrogen and sodium azides and nitride anion. The analyses of the Cr–N, Cr–N–N and N–N BCPs for the compound studied with the B3LYP/6-31G* method are reported and compared with the values obtained for the already mentioned compounds at the same level in Tables 1.3 and 1.4. Two important observations should be done, in one case, the Cr–N BCPs have the topological properties slightly different from those calculated for Na–N BCP of sodium azide, these being the typical properties of the closed-shell interaction ($\rho(r) = 0.04$ a.u., $|\lambda 1|/\lambda 3 < 1$, and $\nabla^2\rho(r) = 0.30$ a.u.) [29–36]. Thus, the structure for sodium azide is ionic with a calculated Na–N distance of 2.05 Å, this being lower than the experimental value reported by Pringles et al. of 2.50 Å [12]. The topological parameters for the compound, such as the Cr–N BCPs is closer to the value obtained for chromyl chloride probably because the electronegativity of the Cl atom (3.16) is similar to that of the N atom (3.04) and for this, the calculated Cr–N distance is closer to the corresponding

Table 1.3 Analysis of Cr–N bond critical points in chromyl azide compared with similar molecules *

B3LYP/6-31G*method

Parameter	$CrO_2(N_3)_2^a$		$CrO_2Cl_2^a$	$CrO_2F_2^a$	NaN_3^a	HN_3^a
	Cr_1-N_4	Cr_1-N_7	Cr–Cl	Cr–F	Na–N	H–N
$\rho(r)$	0.14252	0.14252	0.10851	0.17525	0.04117	0.31882
$\nabla^2\rho(r)$	0.52703	0.52436	0.29701	1.12065	0.31484	−1.57878
$\lambda1$	−0.26409	−0.26513	−0.15099	−0.36390	−0.05346	−1.19248
$\lambda2$	−0.22904	−0.22844	−0.14685	−0.36085	−0.05344	−1.15997
$\lambda3$	1.02016	1.01793	0.59486	1.81540	0.42175	0.77367
$|\lambda1|/\lambda3$	0.25887	0.26045	0.25382	0.20045	0.12675	1.54133
M-X (Å)[b]	1.835	1.835	2.126	1.698	2.045	1.023
[c]X^E	Cr = 1.66	N = 3.04	Cl = 3.16	F = 3.98	Na = 0.93	H = 2.20
M^d (uma)	Cr = 51.9960	N = 14.0067	Cl = 35.453	F = 18.9980	Na = 22.9897	H = 1.0079

* The quantities are in atomic units
[a] This work
[b] (M=Cr, Na; X=N, Cl, F)
[c] X^E : Electronegativity of X

experimental value reported for CrO_2Cl_2 [18]. The interaction H–N bond has the characteristic of the shared interaction, which is the value of the electron density at the bond critical point relatively high and the Laplacian parameter of the charge density as negative indicating that the electronic charge is concentrated in the internuclear region. This value of the Laplacian parameter of the charge density is reasonable compared with the reported value of −1.096 a.u. for the C–H BCP in the VMe_5 compound [29]. The other important observation is that the topological properties of the Cr–N–N and terminal N–N BCPs have different values in the chromyl, hydrogen and sodium azide and nitride anion compounds (Table 1.4).

In the azide compounds, the terminal N–N BCPs have higher $\rho(r)$ values and also the Laplacian parameter at these points have negative values. Only in the nitride anion the properties of the two N–N BCPs have similar values since both bonds are of the same nature (Table 1.4). Moreover, the pathlength of the BCP in hydrogen and sodium azides are different from each other; thus, in the HN_3 compound the distance from BCP to H atom is 0.435, in sodium azide the pathlength is 1.7723, while that value in chromyl azide is closer to sodium azide, it being 1.7557.

The bond orders, expressed by Wiberg's indexes for chromyl azide using a B3LYP/6-31G* calculation are shown in Table 1.5, while those corresponding to chromyl chloride and fluoride compounds are shown in Table 1.6. In the C_2 structure of chromyl azide and C_{2v} for the chloride and fluoride compounds, the chromium atom forms four bonds where the values for chromyl azide are the following: two Cr=O bonds (bond order 1.8351) and two Cr–N (bond order 0.7223), while the bond orders for chromyl chloride (Cr=O 1.9148, Cr–Cl 0.8196) are higher than those corresponding to chromyl fluoride (Cr=O 1.8952, Cr–F 0.7508). The high value of the H–N bond order (0.8006) for HN_3 (Table 1.7) indicates a covalent bond, while a

Table 1.4 Analysis of N–N bond critical points in chromyl azide compared with similar molecules *

B3LYP/6-31G*method

Parameter	$CrO_2(N_3)_2^a$				NaN_3^a		HN_3^a		$N_3^{-\ a}$
	$Cr-N_4-N_5$	$Cr-N_7-N_8$	N_5-N_6	N_8-N_9	$Na-N-N$	$N-N$	$H-N-N$	$N-N$	$N-N$
$\rho(r)$	0.445926	0.445926	0.55904	0.55904	0.45529	0.52594	0.43200	0.56121	0.48725
$\nabla^2\rho(r)$	−0.95596	−0.95596	−1.56165	−1.56165	−1.06684	−1.33520	−0.87501	−1.56850	−1.10488
$\lambda 1$	−0.87795	−0.87795	−1.12665	−1.12665	−0.89100	−0.99151	−0.89553	−1.14940	−0.88568
$\lambda 2$	−0.84914	−0.84914	−1.10776	−1.10776	−0.89088	−0.99150	−0.72022	−1.08079	−0.88568
$\lambda 3$	0.77114	0.77114	0.67277	0.67277	0.71504	0.64781	0.74075	0.66169	0.66658
$|\lambda 1|/\lambda 3$	1.1385	1.1385	1.67464	1.67464	1.24608	1.53055	1.20895	1.73707	1.32869

* The quantities are in atomic units

[a] This work

Table 1.5 Wiberg Index bond matrix of chromyl azide

B3LYP/6−31G* method

$CrO_2(N_3)_2^a$

Atoms		1	2	3	4	5	6	7	8	9
1.	Cr	0.0000	1.8351	1.8351	0.7223	0.0117	0.1274	0.7223	0.0117	0.1274
2.	O	1.8351	0.0000	0.2487	0.0822	0.0059	0.0209	0.0966	0.0064	0.0401
3.	O	1.8351	0.2487	0.0000	0.0966	0.0064	0.0401	0.0822	0.0059	0.0209
4.	N	0.7223	0.0822	0.0966	0.0000	1.5381	0.4972	0.0402	0.0038	0.0227
5.	N	0.0117	0.0059	0.0064	1.5381	0.0000	2.3161	0.0038	0.0001	0.0017
6.	N	0.1274	0.0209	0.0401	0.4972	2.3161	0.0000	0.0227	0.0017	0.0185
7.	N	0.7223	0.0966	0.0822	0.0402	0.0038	0.0227	0.0000	1.5381	0.4972
8.	N	0.0117	0.0064	0.0059	0.0038	0.0001	0.0017	1.5381	0.0000	2.3161
9.	N	0.1274	0.0401	0.0209	0.0227	0.0017	0.0185	0.4972	2.3161	0.0000

[a] This work

Table 1.6 Wiberg Index bond matrix of chromyl azide

B3LYP/6-31G* method

$CrO_2Cl_2^a$

	Atoms	1	2	3	4	5
1.	Cr	0.0000	1.9148	1.9148	0.8196	0.8196
2.	O	1.9148	0.0000	0.2300	0.1285	0.1285
3.	O	1.9148	0.2300	0.0000	0.1285	0.1285
4.	Cl	0.8196	0.1285	0.1285	0.0000	0.0548
5.	Cl	0.8196	0.1285	0.1285	0.0548	0.0000

$CrO_2F_2^a$

	Atoms	1	2	3	4	5
1.	Cr	0.0000	1.8952	1.8952	0.7508	0.7508
2.	O	1.8952	0.0000	0.2492	0.1151	0.1151
3.	O	1.8952	0.2492	0.0000	0.1151	0.1151
4.	F	0.7508	0.1151	0.1151	0.0000	0.0411
5.	F	0.7508	0.1151	0.1151	0.0411	0.0000

[a] This work

low value of Na–N bond order of 0.1297 for NaN_3 shows an ionic nature, as it was observed in Sr (0.14) and Ba (0.12) azides [7].

In this case, the values for chromyl azide are similar to those obtained for chromyl fluoride in accordance with the geometrical parameters previously analyzed. The Cr–N, Cr–N–N and terminal N–N bond orders for chromyl azide are similar to the corresponding values of the HN_3 compound and for this, those bond orders have values characteristic of a slightly covalent bond. In spite of the density of the H–N BCP being higher than the one corresponding to Cr–N BCPs, the bond orders are similar in both cases and, for this reason, the Cr–N bonds in chromyl azide have a partial ionic character. Moreover, the value observed in the natural

Table 1.7 Wiberg Index bond matrix of sodium, hydrogen azide, and nitride anion

		B3LYP/6-31G* method			
		NaN$_3^a$			
	Atoms	1	2	3	4
1.	Na	0.0000	0.1297	0.0048	0.0356
2.	N	0.1297	0.0000	1.7309	0.7767
3.	N	0.0048	1.7309	0.0000	2.1443
4.	N	0.0356	0.7767	2.1443	0.0000
		HN$_3^a$			
	Atoms	1	2	3	4
1.	H	0.0000	0.8006	0.0049	0.0416
2.	N	0.8006	0.0000	1.5289	0.6132
3.	N	0.0049	1.5289	0.0000	2.3499
4.	N	0.0416	0.6132	2.3499	0.0000
		B3LYP/6-31G* method		
		N$_3^-$		
	Atoms	1	2	3
1.	N	0.0000	1.9304	0.8642
2.	N	1.9304	0.0000	1.9304
3.	N	0.8642	1.9304	0.0000

[a] This work

charge of the internal N atom (Cr–N) of −0.51463 reveals that the bond nature in this compound is partial ionic because that value is lower than the natural charge value observed in the N atom linked to the Na atom (−0.82083) (Table 1.8).

1.3 Vibrational Analysis

The structure of chromyl azide has a C_2 symmetry and 21 active vibrational normal modes in the infrared and Raman spectra classified as: 11 A + 10 B. Taking into account that the theoretical infrared spectra for chromyl chloride and sodium azide compounds are similar to the ones corresponding to chromyl azide, as shown in Fig. 1.2, the experimental bands of those compounds were proposed as experimental bands for chromyl azide in order to effect a complete assignment of the compound by means of the SQMFF method.

With the exception of the bands in the region between 110 and 50 cm^{-1} the calculated frequencies were considered as experimental because they could not be seen in the vibrational spectrum of those similar molecules. The calculated frequencies and the assignment for chromyl azide are given in Table 1.9. The theoretical calculations reproduce the proposed normal frequencies for the compound with an initial value of mean square root deviations (RMSD) of 61.1 cm^{-1} using

Table 1.8 Analysis of Cr–N bond critical points in chromyl azide compared with similar molecules

B3LYP/6-31G*method[a]

CrO2(N3)2			CrO2Cl2			CrO2F2[a]			Na(N3)		
Atoms	NPA	Wiberg index	Atoms	NPA	Wiberg index	Atoms	NPA	Wiberg index	Atoms	NPA	Wiberg index
Cr1	1.43473	0.0000	Cr1	1.33662	1.57420	Cr1	0.0000	0.0000	Na1	0.92049	0.0000
O2	−0.40080	1.8351	O2	−0.33956	−0.36486	O2	1.9148	1.8952	N2	−0.82083	0.1297
O3	−0.40080	1.8351	O3	−0.33956	−0.36486	O3	1.9148	1.8952	N3	0.18160	0.0048
N4	−0.51463	0.7223	Cl4	−0.32875	−0.42224	F4	0.8196	0.7508	N4	−0.28126	0.0356
N5	0.19385	0.0117	Cl5	−0.32875	−0.42224	F5	0.8196	0.7508			
N6	0.00422	0.1274									
N7	−0.51463	0.7223									
N8	0.19385	0.0117									
N9	0.00422	0.1274									

The quantities are in atomic units

[a] This work

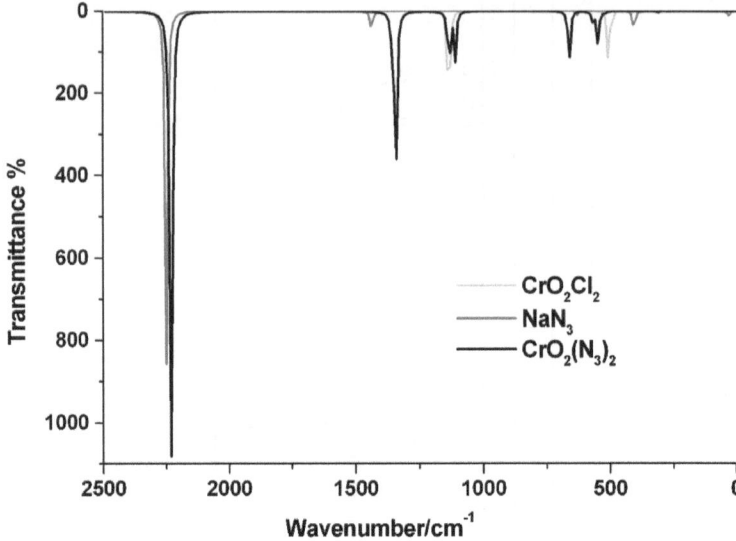

Fig. 1.2 Theoretical Infrared spectrum of $CrO_2(N_3)_2$ at B3LYP/6-31G* theory levels compared with similar compounds

the B3LYP/6-31G* method. When the SQMFF method is applied, the final RMSD decreases significantly up to 2.6 cm^{-1}. Vibrational assignments were made on the basis of potential energy distributions (PED) in terms of symmetry coordinates and by comparison with molecules containing similar groups such as the chromyl chloride [17] and fluoride [20] compounds, hydrogen [37] and sodium [12, 13] azides and nitride anion [12]. Table 1.10 shows the calculated harmonic frequencies for chromyl azide using the B3LYP method with the 6-31G* basis set compared with experimental values of similar molecules. In general, the theoretical infrared and Raman spectra of the chromyl azide agrees with the proposed experimental spectrum (see Fig. 1.2). Some vibration modes of different symmetries may be seen mixed among them because the frequencies are approximately the same. A discussion of the assignment of the most important groups for the compounds studied is presented below (see Table 1.9).

1.4 Chromyl Groups

In chromyl compounds these modes appear in the 1,050–900 cm^{-1} region, i.e., in $CrO_2(ClO_4)_2$, they appear at 990 and 980 cm^{-1} [38, 39], while in $CrO_2(SO_3F)_2$ they do so at 1,061 and 1,020 cm^{-1} [40, 41]. The frequencies predicted for the vibrational modes of chromyl azide show that the anti-symmetric and symmetric Cr=O

Table 1.9 Calculated and experimental predict frequencies (cm^{-1}), potential energy distribution, and assignment for chromyl azide

Modes	IR[a]	Calculated[b]	SQM[c]	Assignment, PED ($\geq$10 %)
A symmetry				
1	2140	2242	2153	νip (N=N)$_e$ (77), νip (N=N)$_i$ (23)
2	1309	1356	1304	νip (N=N)$_i$ (75), νip (N=N)$_e$ (15)
3	991	1110	987	ν_sCr=O (96)
4	607	661	602	δN–N–N op (60), δCr–N–N op (15)
5	534	566	530	τCr–N–N–N ip (87)
6	432	518	447	ν_aCr–N (62), wagCrO$_2$ (20)
7	358	382	353	ν_sCr–N (71), δCrO$_2$ (14), δN–N–N op (12)
8	220	262	230	ρCrO$_2$ (88)
9	214	234	219	δN–Cr–N (75), τO–Cr–N–N op (18)
10	110	110	107	τO–Cr–N–N ip (51), δCr–N–N op (20), wagCrO$_2$ (10)
11	37	37	42	τO–Cr–N–N op (73)
B symmetry				
12	2140	2227	2138	νop (N=N)$_e$ (78), νop (N=N)$_i$ (22)
13	1309	1343	1291	νop (N=N)$_i$ (78), νop (N=N)$_e$ (13)
14	1002	1134	1006	ν_aCr=O (98)
15	607	663	616	δN–N–N ip (72), δCr–N–N ip (10)
16	534	571	538	τCr–N–N–N op (86)
17	473	548	460	δCrO$_2$ (72), ν_sCr–N (13)
18	268	313	274	wagCrO$_2$ (60), τCrO$_2$ (14), ν_aCr–N (13)
19	257	271	241	τCrO$_2$ (57), wagCrO$_2$ (18), τO–Cr–N–N ip (11)
20	106	106	99	δCr–N–N ip (77)
21	66	66	67	δCr–N–N op (43), τO–Cr–N–N ip (34), τCrO$_2$ (11)
RMSD (cm^{-1})		61.12	2.60	

[a] This work
[b] DFT B3LYP/6-31G*
[c] From scaled quantum mechanics force field

stretchings are split by more than 20 cm^{-1}, indicating a notable contribution of the central Cr atom in these vibrations. The antisymmetric and symmetric Cr=O stretching modes were calculated in the infrared spectrum at 1,134 and 1,110 cm^{-1}, respectively, the last band being more intense in the Raman spectrum. As in similar molecules [39, 41–43], these modes are calculated uncoupled with other modes. Taking into account the difference between those modes, they are assigned, as in CrO$_2$Cl$_2$, at 1,002 and 991 cm^{-1}, thus, the scaled frequencies for both modes decrease up to 1,009 and 984 cm^{-1}, respectively, with a difference between them of 25 cm^{-1}. Table 1.9 shows for chromyl azide that the unscaled DFT frequencies for the symmetric Cr=O stretching mode are higher than the frequencies of the antisymmetric Cr=O stretchings. The antisymmetric and symmetric Cr–N stretchings are also split by more than 130 cm^{-1}, indicating a strong contribution of the central Cr atom in these vibrations. In this case, the theoretical calculations predict these modes with a higher PED (79 %) value for the antisymmetric mode in reference to the symmetric mode (48 %). For this reason, when they are assigned at 432 and

358 cm^{-1} calculated at 406 cm^{-1} for the Na–N stretching of the NaN$_3$ compound using the same method, then, the difference between both scaled modes decreases at 74 cm^{-1}. The CrO$_2$ bending mode can be seen in CrO$_2$F$_2$ at 403 cm^{-1} while in CrO$_2$Cl$_2$ it is at 356 cm^{-1} [17, 20]. In this chapter, the B3LYP/6-31G* method calculates the CrO$_2$ bending at 548 cm^{-1} with 48 % of PED contribution and at 382 cm^{-1} with 41 % of PED contribution When this mode is assigned at 473 cm^{-1} as the theoretical value obtained for CrO$_2$Cl$_2$ using the same level, the scaled frequency decreases up to 460 cm^{-1}. For chromyl compounds, the calculations normally predict the wagging, rocking, and twisting modes of the CrO$_2$ group in the low frequency region and couples with other modes of the azide groups. The wagging CrO$_2$ mode is calculated at a higher frequency and with a PED contribution (42 %) lower than the rocking mode (86 %), as it was observed in similar compounds [39, 41, 43]. For these observations, those modes are assigned, respectively, at 268 and 220 cm^{-1}. As indicated by the PED values, the CrO$_2$ twisting mode is calculated strongly coupled with vibrations of the same group and the azide groups (Table 1.9). This mode could be assigned at 257 cm^{-1} because it appears with a higher PED value. The N–Cr–N bending mode is calculated at 234 cm^{-1} with 75 % of PED contribution, while the Cl–Cr–Cl bending is experimentally observed for CrO$_2$Cl$_2$ at 214 cm^{-1}. As the N–Cr–N bond angle value (109.3°) is near the one observed for Cl–Cr–Cl bond angle of chromyl chloride (113.2°), for this reason, the N–Cr–N bending mode is also assigned at the same frequency, which is at 214 cm^{-1}, then the scaled value increases at 219 cm^{-1}.

1.5 Azide Groups

In this chapter, the external or terminal N=N in-phase and out-of-phase symmetric stretching modes of the N=N group are calculated by the B3LYP/6-31G* method at 2,242 and 2,227 cm^{-1}, respectively. The frequencies predicted for these vibrational modes show that these modes are split by about 16 cm^{-1}. They are calculated slightly coupled with the internal N=N antisymmetric and symmetric stretching modes (N=Ni) and they are assigned at 2,140 cm^{-1}, the same as for the N=Ne stretching of NaN$_3$ [8]. Then the scaled frequencies of both modes decrease until there is a difference of 15 cm^{-1} between them. On the other hand, the internal N=N symmetric and antisymmetric stretching modes (N=Ni) are calculated at 1,356 and 1,343 cm^{-1}, respectively, and they are assigned at 1,309 cm^{-1}, as it could also be seen in the corresponding IR spectrum of NaN$_3$ for the N=Ni stretching (Fig. 1.2). Experimentally, for the hydrogen azide compound, the difference between the N=Ni and N=Ne stretching modes is 874 cm$^{-1,}$ while for sodium azide it is 812 cm^{-1}, for chromyl azide the calculated values using B3LYP/6-31G* level vary from 900 to 884 cm^{-1}. The difference between the scaled frequencies of both modes results from 12 cm^{-1} in agreement with the value obtained by the calculations (13 cm^{-1}). The N=N=N out-of-phase and in-phase deformation modes are calculated coupled with the Cr–N–N out-of-phase and in-phase deformation modes. For chromyl azide, the

Table 1.10 Calculated frequencies (cm^{-1}) for chromyl and sodium azide compared with experimental values of similar molecules

CrO$_2$(N$_3$)$_2$[a]	Assignment	CrO$_2$Cl$_2$[b]	CrO$_2$F$_2$[c]	H(N$_3$) [d]	Na(N$_3$) [a]
2242	νip (N=N)$_e$			2140	2248
2227	νop (N=N)$_e$				
1356	νip (N=N)$_i$			1151	1436
1343	νop (N=N)$_i$				
1134	ν_aCr=O	1002	1045		
1110	ν_sCr=O	991	995		
663	δN–N–N ip			607	631
661	δN–N–N op			534	631
571	τCr–N–N op				
566	τCr–N–N ip				
548	δCrO$_2$	356	403		
518	ν_aCr–N	503	770		406
382	ν_sCr–N	470	708		
313	wagCrO$_2$				
271	τCrO$_2$	224	275		
262	ρCrO$_2$				
234	δN–Cr–N	234			
110	τN–N–N ip				
106	δCr–N–N ip				32
66	δCr–N–N op				
37	τN–N–N op				

Abbrevitations ν stretching, δ deformation, ρ rocking, *wag* wagging, τw torsion, *a* antisymmetric, *s* symmetric

op out of phase, *ip* in phase, *vs* very strong, *s* strong, *m* medium, *w* weak, *sh* shoulder

[a] This work B3LYP/6-31G*
[b] Ref [20]
[c] Ref [17]
[d] Ref [25]

former are calculated, respectively, at 661 and 663 cm^{-1} while for NaN$_3$, these modes are both calculated at the same frequency (631 cm^{-1}) by using the B3LYP/6-31G* method and for this reason, they were assigned at 607 cm^{-1} with values of the scaled frequencies at 602 and 616 cm^{-1}, respectively. In the region of low frequencies, the vibration modes normally expected for the azide groups are the Cr–N–N in-phase and out-of-phase deformation modes and the O–Cr–N–N in-phase and out-of-phase torsion modes as predicted by calculations. All modes, as shown in Table 1.9, appear coupled with each other and with other modes of the chromyl group. The modes are assigned at the same frequencies as predicted by calculations.

Table 1.11 Definition of natural internal coordinates for chromyl azide

A symmetry	
$S_1 = s\,(5\text{-}6) + s\,(8\text{-}9)$	$\nu ip\,(N{=}N)_e$
$S_2 = s\,(4\text{-}5) + s\,(7\text{-}8)$	$\nu ip\,(N{=}N)_i$
$S_3 = q\,(1\text{-}2) + q\,(1\text{-}3)$	$\nu_s Cr{=}O$
$S_4 = \phi\,(4\text{-}5\text{-}6) - \phi\,(7\text{-}8\text{-}9)$	$\delta N\text{-}N\text{-}N$ op
$S_5 = \tau\,(1\text{-}4\text{-}5\text{-}6) + \tau\,(1\text{-}7\text{-}8\text{-}9)$	$\tau Cr\text{-}N\text{-}N\text{-}N$ ip
$S_6 = r\,(1\text{-}4) - r\,(1\text{-}7)$	$\nu_a Cr\text{-}N$
$S_7 = r\,(1\text{-}4) + r\,(1\text{-}7)$	$\nu_s Cr\text{-}N$
$S_8 = \psi\,(3\text{-}1\text{-}6) + \psi\,(3\text{-}1\text{-}4) - \psi\,(2\text{-}1\text{-}4) - \psi\,(2\text{-}1\text{-}6)$	$\rho Cr O_2$
$S_9 = \alpha\,(9\text{-}10\text{-}12)$	$\delta N\text{-}Cr\text{-}N$
$S_{10} = \tau\,(3\text{-}1\text{-}4\text{-}11) + \tau\,(2\text{-}1\text{-}4\text{-}11) - \tau\,(2\text{-}1\text{-}6\text{-}10) - \tau\,(3\text{-}1\text{-}6\text{-}10)$	$\tau O\text{-}Cr\text{-}N\text{-}N$ ip
$S_{11} = \tau\,(3\text{-}1\text{-}4\text{-}11) + \tau\,(2\text{-}1\text{-}4\text{-}11) - \tau\,(2\text{-}1\text{-}6\text{-}10) - \tau\,(3\text{-}1\text{-}6\text{-}10)$	$\tau O\text{-}Cr\text{-}N\text{-}N$ op
B Symmetry	
$S_{12} = s\,(5\text{-}6) - s\,(8\text{-}9)$	$\nu op\,(N{=}N)_e$
$S_{13} = s\,(4\text{-}5) - s\,(7\text{-}8)$	$\nu op\,(N{=}N)_i$
$S_{14} = q\,(1\text{-}2) - q\,(1\text{-}3)$	$\nu_a Cr{=}O$
$S_{15} = \phi\,(4\text{-}5\text{-}6) + \phi\,(7\text{-}8\text{-}9)$	$\delta N\text{-}N\text{-}N$ ip
$S_{16} = \tau\,(1\text{-}4\text{-}5\text{-}6) - \tau\,(1\text{-}7\text{-}8\text{-}9)$	$\tau Cr\text{-}N\text{-}N\text{-}N$ op
$S_{17} = \theta\,(2\text{-}1\text{-}3)$	$\delta Cr O_2$
$S_{18} = \psi\,(2\text{-}1\text{-}4) + \psi\,(3\text{-}1\text{-}4) - \psi\,(2\text{-}1\text{-}6) - \psi\,(3\text{-}1\text{-}6)$	$wag Cr O_2$
$S_{19} = \psi\,(3\text{-}1\text{-}4) + \psi\,(2\text{-}1\text{-}6) - \psi\,(3\text{-}1\text{-}6) - \psi\,(2\text{-}1\text{-}4)$	$\tau Cr O_2$
$S_{20} = \alpha\,(1\text{-}4\text{-}5) + \alpha\,(1\text{-}7\text{-}8)$	$\delta Cr\text{-}N\text{-}N$ ip
$S_{21} = \beta\,(1\text{-}4\text{-}5) - \alpha\,(1\text{-}7\text{-}8)$	$\delta Cr\text{-}N\text{-}N$ op

$q = Cr{=}O$ bond distance, $r = Cr\text{-}N$ bond distance, $s = N\text{-}N$ bond distance, $\theta = O{=}Cr{=}O$ bond angle, $\alpha = N\text{-}Cr\text{-}N$ bond angle, $\phi = N\text{-}N\text{-}N$ bond angle, $\psi = O{=}Cr\text{-}N$ bond angles, $\beta = Cr\text{-}N\text{-}N$ bond angle

Abbrevitations ν stretching, δ deformation, ρ in the plane bending or rocking, *wag* out of plane bending or wagging, τ torsion, *a* antisymmetric, *s* symmetric, *ip* in phase, *op* out of phase

Table 1.12 Scale factors for the force field of chromyl azide

Coordinates	B3LYP/6-31G*			
	$CrO_2(N_3)_2^a$	HN_3^a	NaN_3^a	N_3^-
ν (N=N)	0.922	0.870	0.882	0.962
ν (Cr=O)	0.787	–	–	–
ν (Cr–N)	0.962	0.925	0.958	–
δ (O=Cr=O)	0.874	–	–	–
δ (N–Cr–N)	0.823	–	–	–
δ (Cr–N–N)	0.823	0.834	0.999	–
δ (N–N–N)	0.614	1.021	1.000	0.999
ρ (CrO$_2$)	0.769	–	–	–
Wagg (CrO$_2$)	0.769	–	–	–
τw (CrO$_2$)	0.769	–	–	–

ν stretching, δ deformation, ρ rocking, *wag* wagging, τw torsion
[a] This work

Table 1.13 Main force constant for chromyl azide

Force constants[#]	B3LYP					
	6-31G*	Lanl2dz		6-31G*		
	$CrO_2(N_3)_2^a$	$CrO_2Cl_2^b$	$CrO_2F_2^b$	$H(N_3)^a$	$Na(N_3)^a$	N_3^{-a}
f (Cr=O)	7.04	7.122	7.443	–	–	–
f (M-X)	3.19	2.857	4.491	6.17	1.45	–
f (O–Cr–O)	1.25	0.938	1.110	–	–	–
f (X-Cr-X)	0.59	0.458	0.460	–	–	–
f (O-M-X)	0.58	0.553	0.656	–	–	–
f (N–N)e	17.29	–	–	16.74	14.54	14.79
f (N–N)i	9.93	–	–	9.27	12.21	12.37
f (N–N–N)	0.40	–	–	0.64	0.80	0.87

Units: mdyn $Å^{-1}$ for stretching and mdyn $Å$ rad^{-2} for bond deformation
[a] This work
[b] Ref [48]
[#] (M=Cr, Na; X=N, Cl, F)

1.6 Force Field

Having a proposed assignment for chromyl azide, the corresponding force constants were estimated using the scaling procedure of Pulay et al. [44], as mentioned before. The harmonic force fields in Cartesian coordinates were turned into the local symmetry or "natural" coordinates proposed by Fogarasi et al. [45], as defined in Table 1.11 (See Fig. 1.1.).

The scaling factors affecting the main force constants were subsequently calculated by an iterative procedure [46, 47] to have the best possible fit between the ones proposed as experimental and theoretical frequencies. The resulting factors are shown in Table 1.12. The experimental proposed frequencies and potential energy distribution obtained for chromyl azide appear together with the values reached for the corresponding RMSD values in Table 1.9. The main force constants appearing in Table 1.13 were compared with the calculated force constants at the B3LYP/Lanl2DZ for CrO_2Cl_2 and CrO_2F_2 [48] and with those calculated in this work at B3LYP/6-31G* level for hydrogen, sodium azide, and nitride ion. In CrO_2F_2 and CrO_2Cl_2, the scaled GVFF force constants (B3LYP/Lanl2DZ method) for the Cr=O stretchings are, respectively, 7.443 and 7.122 mdyn $Å^{-1}$, while the corresponding force constants for the O=Cr=O deformations are 1.110 and 0.938 mdyn $Å$ rad^{-2} [19].

This difference in the force constant values in reference to chromyl azide cannot be attributed to geometrical parameters since they are practically the same in the three compounds. The corresponding values are low in CrO_2F_2 and CrO_2Cl_2 because the scaled Cr=O frequencies are higher than the values for $CrO_2(N_3)_2$ (1,006 and 987 cm^{-1}) as shown in Table 1.9. Moreover, the calculated Cr=O stretching force constants increase according to the electronegativity of the other atoms linked to Cr atom increase, it is, $f(Cr=O)CrO_2(N_3)_2 < f(Cr=O)CrO_2Cl_2 < f(Cr=O)CrO_2F_2$. The low values of the force constants of O=Cr=O deformations in CrO_2F_2 and CrO_2Cl_2 in comparison with the ones corresponding to $CrO_2(N_3)_2$ can also be attributed to the scaled O=Cr=O frequencies because they are high in the last compound as it was observed in $CrO_2(NO_3)_2$ [43]. As expected, the M-X and X-M-X force constants change with the corresponding geometrical parameters, the f(M-X) value for CrO_2F_2 (4.491 mdyn Å^{-1}) being high with a low M-X distance of 1.716 Å. On the other hand, the f(X-Cr-X) force constant value is higher for chromyl azide (0.59 mdyn Å rad^{-2}) because the X-Cr-X bond angle of 109.3° is lower in the remaining compounds. The value of 3.19 mdyn Å^{-1} for the Cr–N force constant suggests that the Cr atom is bonded to the azide groups in chromyl azide by means of ionic bonds since that value is slightly higher than the one corresponding to CrO_2ClN_3 compound (2.32 mdyn Å^{-1}) where the Cr–N bond is partly ionic [1]. The f(N–N)e and f(N–N)i stretchings and f(N–N–N) force constants for chromyl azide are compared with the values obtained for similar molecules in Table 1.13. The obtained values are close to the calculated values for the HN_3 compound indicating a partial covalent contribution of these bonds at the Cr atom. Note that when the azide group is linked to ionic atoms (as Na) the f(N–N)e and f(N–N)i force constant values are closer to each other.

1.7 Electronic Spectrum and Energy Gap

The UV–vis spectrum of chromyl azide in CCl_4 solution previously reported shows only two definite bands, one with a maximum at 545 and the other at 459 nm [15]. The former probably has its origin in the azide groups and the latter can be assigned to chromyl groups as it was observed in the electronic spectra of chromyl compounds [42, 49]. In the theoretical B3LYP/6-31G* UV–visible spectrum, the first band appears as a triplet of B symmetry at 491.17 nm and with a low energy (2.5242 eV), while the second band is observed as a singlet of B symmetry at 459.66 nm and with an energy of 2.6973 eV. The slight difference in the energy of the first observed band, in relation to the calculation, can be attributed to the solvent effect because in this last case it was not taken into account.

Table 1.14 HOMO and LUMO energy (au.) and HOMO–LUMO (GAP) for chromyl azide compared with similar molecules [*]

B3LYP/6-31G*method

Orbital	$CrO_2(N_3)_2^a$	$CrO_2Cl_2^a$	$CrO_2F_2^a$	$Na(N_3)^a$	HN_3^a	N_3^{-a}
HOMO	−0.30081	−0.34956	−0.36485	−0.17062	−0.27585	0.04241
LUMO	−0.17296	−0.19610	−0.18440	−0.07849	−0.03549	0.31859
GAP	0.12785	0.15346	0.18045	0.24911	0.24036	0.27618
GAP (eV)	2.95	3.54	4.16	5.74	5.54	6.37

[*] The quantities are in atomic units
[a] This work

In addition, for chromyl azide the magnitude of the band gap was investigated because the azide compounds are unstable and break down under the action of light and heat. The magnitude of the band gap for the compound compared with other azides can be seen in Table 1.14. In general, ionic azides are stable (insulators as $Sr(N_3)_2$ and $Ba(N_3)_2$ with values of 3.708 and 3.649 eV, respectively), while heavy-metal azides explode on provocation, and covalent azides explode spontaneously [7]. For chromyl azide, the 2.95 eV value indicates that the compound is more energetic than sodium azide.

1.8 Conclusions

In this chapter an approximate normal coordinate analysis was proposed for chromyl azide at the vibrational spectra and a complete assignment of the 21 normal modes of vibration is reported.

An SQM force field was obtained for chromyl azide after adjusting the theoretically obtained force constants in order to minimize the difference between the proposed and calculated frequencies.

For the chromyl azide it is demonstrated that a DFT molecular force field computed using B3LYP/6-31G* is well represented [50–53].

The NBO [54], AIM, and energy gap analyses confirm a slightly ionic character of the Cr atom bonded to azide groups in chromyl azide.

Acknowledgments This work was subsidized with grants from CIUNT (Consejo de Investigaciones, Universidad Nacional de Tucumán). The author thanks Prof. Tom Sundius for his permission to use MOLVIB.

References

1. H.D. Fair, R.F. Walker, *Physics and Chemistry of the Inorganic Azides (Vol. 1) and the Technology of the Inorganic Azides (Vol. 2)*
2. R. Vince, S. Daluge, J. Med. Chem. **17**(6), 578 (1974)
3. M.R. Darvich, A. Olyai, A. Shabanikia, D. Farkhani, Iran J. Chem. Chem. Eng. **22**(2), 49–53 (2003)
4. K.A. Jensen, C. Pedersen, Acta Chem. Scand. **15**, 1104–1108 (1961)
5. R. Sreekumar, R. Padmakumar, P. Rugmini, J. Chem. Commun. **12**, 1133 (1997)
6. E.H. Younk, A. Barry Kunz, Int. J. Quantum Chem. **63**(3), 615–621 (1998)
7. W. Zhu, X. Xu, H. Xiao, J. Phys. Chem. Solids **68**(9), 1762–1769 (2007)
8. R.T.M. Fraser, Anal. Chem. **31**(9), 1602–1603 (1959)
9. J.I. Bryant, J. Chem. Phys. **42**(7), 2270–2273 (1965)
10. N. Chalermsan, P. Vijchulata, P. Chirattanayuth, S. Sintuwanit, S. Surapat, A. Engkagul, Kasetsart J. (Nat. Sci.) 38, 38 (2004)
11. E. Qamirani, M.H. Razavi, Wu Xin; J. M. Davis, L. Kuo, W. T. Hein, *Am. J.Physiology*, **2006**, 290(4), H1617
12. G.E. Pringles, D.E. Noakes, Acta Crystallogr. **B24**, 262 (1968)

13. E.D. Stevens, H. Hope, Acta Crystallogr. **A33**, 723–729 (1977)
14. S.R. Aghdaee, A.I.M. Rae, Acta Crystallogr. **B40**, 214–218 (1984)
15. H.L. Krauss, F. Schwarzbach, Chem. Ber. **94**, 1205 (1961)
16. H.L. Krauss, K. Stark, Z. Naturforsch. **17**, 345 (1962)
17. H. Stammreich, K. Kawai, Y. Tavares, Spectrochim. Acta **75**, 438 (1959)
18. C.J. Marsden, L. Hedberg, K. Hedberg, Inorg. Chem. **21**, 1113 (1982)
19. R.J. French, L. Hedberg, K. Hedberg, G.L. Gard, B.M. Johnson, Inorg. Chem. **22**, 892 (1982)
20. S.D. Brown, G.L. Gard, T.M. Lohehr, J. Chem. Phys. **21**, 1115 (1982)
21. A.E. Reed, L.A. Curtis, F. Weinhold, Chem. Rev. **88**(6), 899 (1988)
22. J.P. Foster, F. Weinhold, J. Am. Chem. Soc. **102**, 7211 (1980)
23. A.E. Reed, F. Weinhold, J. Chem. Phys. **83**, 1736 (1985)
24. R.F.W. Bader, *Atoms in Molecules, A Quantum Theory* (Oxford University Press, Oxford, 1990). ISBN 0198558651
25. Tables of inter-atomic distances and configurations in molecules and anions, Edit. L. E. Sutton, **1958**, p. 49
26. R.J. Gillespie (ed.), *Molecular Geometry* (Van Nostrand-Reinhold, London, 1972)
27. R.J. Gillespie, I. Bytheway, T.H. Tang, R.F.W. Bader, Inorg. Chem. **35**, 3954 (1996)
28. F. Biegler-Köning, J. Schönbohm, D. Bayles, AIM2000; a program to analyze and visualize atoms in molecules. J. Comput. Chem. **22**, 545 (2001)
29. S. Wojtulewski, S.J. Grabowski, J. Mol. Struct. **621**, 285 (2003)
30. S.J. Grabowski, Monat. für Chem. **133**, 1373 (2002)
31. R.F.W. Bader, J. Phys. Chem. A **102**, 7314 (1998)
32. P.L.A. Popelier, J. Phys. Chem. A **1998**, 102 (1873)
33. U. Koch, P.L.A. Popelier, J. Phys. Chem. **99**, 9747 (1995)
34. G.L. Sosa, N. Peruchena, R.H. Contreras, E.A. Castro, J. Mol. Struct. (THEOCHEM) **401**, 77 (1997)
35. G.L. Sosa, N. Peruchena, R.H. Contreras, E.A. Castro, J. Mol. Struct. (THEOCHEM) **577**, 219 (2002)
36. S. Wojtulewski, S.J. Grabowski, J. Mol. Struct. **645**, 287 (2003)
37. T. Shimanouchi, Tables of Molecular Vibrational Frequencies, **1977**, p. 1019
38. M. Chaabouni, T. Chausse, J.L. Pascal, J. Potier, J. Chem. Research **5**, 72 (1980)
39. S.A. Brandán, Theoretical vibrational study of the chromyl perchlorate, $CrO_2(ClO_4)_2$. J. Mol. Struc. (THEOCHEM) **908**, 19 (2009)
40. S.D. Brown, G.L. Gard, Inorg. Chem. **12**, 483 (1973)
41. A. Ben Altef, S.A. Brandán, A new vibrational study of chromyl fluorosulphate, $CrO_2(SO_3F)_2$ by using DFT calculations. J. Mol. Struc. **981**, 146 (2010)
42. S.A. Brandán, A. Ben Altef, E.L. Varetti, Spectrochim. Acta **51A**, 669 (1995)
43. S.A. Brandán, M.L. Roldán, C. Socolsky, A. Ben Altef, Spectrochim. Acta, Part A **2008**, 69 (1027)
44. P. Pulay, G. Fogarasi, F. Pang, J.E. Boggs, J. Am. Chem. Soc. **101**(10), 2550 (1979)
45. G. Fogarasi, P. Pulay, in *Vibrational Spectra and Structure*, vol 14, ed. by J.E. Durig (Elsevier, Amsterdam, 1985), p. 125
46. T. Sundius, J. Mol. Struct. **218**, 321 (1990)
47. T. Sundius, MOLVIB: A Program for Harmonic Force Field Calculation, QCPE Program No. 604 (1991)
48. S. Bell, T.J. Dines, J. Phys. Chem. A **104**, 11403 (2000)
49. M. Sowinska, J. Myrczek, A. Bartecki, J. Mol. Struct. **218**, 267 (1990)
50. A.B. Nielsen, A.J. Holder, *GaussView, User's Reference* (GAUSSIAN Inc., Pittsburgh, 2000–2003)
51. Gaussian 03, Revision B.01, M.J. Frisch, G.W. Trucks, H.B. Schlegel, G.E. Scuseria, M.A. Robb, J.R. Cheeseman, J.A. Montgomery Jr., T. Vreven, K.N. Kudin, J.C. Burant, J.M. Millam, S.S. Iyengar, J. Tomasi, V. Barone, B. Mennucci, M. Cossi, G. Scalmani, N. Rega, G.A. Petersson, H. Nakatsuji, M. Hada, M. Ehara, K. Toyota, R. Fukuda, J. Hasegawa, M. Ishida, T. Nakajima, Y. Honda, O. Kitao, H. Nakai, M. Klene, X. Li,

J.E. Knox, H.P. Hratchian, J.B. Cross, C. Adamo, J. Jaramillo, R. Gomperts, R.E. Stratmann, O. Yazyev, A.J. Austin, R. Cammi, C. Pomelli, J.W. Ochterski, P.Y. Ayala, K. Morokuma, G.A. Voth, P. Salvador, J.J. Dannenberg, V.G. Zakrzewski, S. Dapprich, A.D. Daniels, M.C. Strain, O. Farkas, D.K. Malick, A.D. Rabuck, K. Raghavachari, J.B. Foresman, J.V. Ortiz, Q. Cui, A.G. Baboul, S. Clifford, J. Cioslowski, B.B. Stefanov, G. Liu, A. Liashenko, P. Piskorz, I. Komaromi, R.L. Martin, D.J. Fox, T. Keith, M.A. Al-Laham, C.Y. Peng, A. Nanayakkara, M. Challacombe, P.M.W. Gill, B. Johnson, W. Chen, M.W. Wong, C. Gonzalez, J.A. Pople (Gaussian Inc., Pittsburgh, 2003)
52. A.D. Becke, J. Chem. Phys. **98**, 5648 (1993)
53. C. Lee, W. Yang, R.G. Parr, Phys. Rev. **B 37**, 785 (1988)
54. E.D. Glendening, A.E. Reed, J.E. Carpenter, F. Weinhold, NBO Version 3.1

Chapter 2
Structural and Vibrational Study on Chromyl Acetate in Different Media

Abstract In this chapter, the structural and vibrational properties of chromyl acetate in different media were studied using density functional theory (DFT) methods. The initial geometries were fully optimized at different theory levels and the harmonic wavenumbers were evaluated at the same levels. Also, the characteristics and nature of the Cr–O and Cr ← O bonds for the stable structure were studied by means of the natural bond orbital (NBO) study while the topological properties of the electronic charge density were analyzed using the Bader atoms in the molecules theory (AIM). Besides, a complete assignment of all observed bands in the infrared spectrum for the compound was performed combining DFT calculations with Pulay's scaled quantum mechanics force field (SQMFF) methodology.

Keywords Chromyl acetate · Vibrational spectra · Molecular structure · Force field · DFT calculations

2.1 Introduction

In organic chemistry, for a long time the chemical reactivity of chromyl acetate, $CrO_2(CH_3COO)_2$ was broadly studied because it is an effective reagent for the oxidation of numerous compounds, such as hydrocarbons [1, 2], olefins [3, 4], C-25 on cholestane side-chains to the corresponding C-25 hydroxy derivative [5], steroids [6] and indole alkaloids [7] among others, but so far their molecular structure is unknown. In a previous paper we could assign some of the observed bands in the infrared spectrum and for this their vibrational properties remain only partially characterized [8]. The study of compounds that contain chromyl groups [8–11] is interesting because depending of the ligand they can act with different

S. A. Brandán, *A Structural and Vibrational Investigation into Chromylazide, Acetate, Perchlorate, and Thiocyanate Compounds*, SpringerBriefs in Molecular Science, DOI: 10.1007/978-94-007-5754-7_2, © The Author(s) 2013

coordination modes. In the chapter above, the structure and vibrational properties of the chromyl nitrate compound, $CrO_2(NO_3)_2$ were studied by means of the Density functional theory [10]. In this chapter, the geometries and the harmonic vibrational frequencies for chromyl acetate were evaluated at the B3LYP/Lanl2DZ, B3LYP/6-31G*, and B3LYP/6-311++G levels of theory. In this way, with the aid of calculations it was possible the assignment of all the observed bands in the vibrational spectra for the compound taking into account the type of coordination adopted by the acetate groups as monodentate and bidentate. In addition, as the nitrate group, the acetate group is an interesting ligand because it presents different coordination modes: monodentate or bidentate [8–23].

The properties of the $CrO_2(CH_3COO)_2$ compound prepared previously by us were completely different from the relatively volatile compound initially prepared according to *Krauss* [23, 24]. In this chapter a theoretical study of chromyl acetate, $CrO_2(CH_3COO)_2$, was performed in order to study the coordination mode of the acetate groups and carry out its complete assignment using the experimental spectra. The aim of this chapter is to carry out an experimental and theoretical study on this compound with the methods of quantum chemistry in order to have a better understanding of its vibrational properties. A precise knowledge of the normal modes of vibration is expected to provide a foundation for understanding the conformation-sensitive bands in vibrational spectra of this molecule. Here, the normal mode calculations were accomplished using a generalized valence force field (GVFF) and considering the acetate group as monodentate and bidentate ligands. For this purpose, the optimized geometry and frequencies for the normal modes of vibration were calculated first in gas phase and then in carbon tetrachloride (CCl_4) and dimethylsulfoxide (DMSO) solutions because the compound is soluble in all the inert organic solvents. In both cases, there are no publications about experimental or high-level theoretical studies on the geometries and force field of chromyl acetate. Hence, obtaining reliable parameters by theoretical methods is an appealing alternative. The parameters obtained may be used to gain chemical and vibrational insights into related compounds.

The election of the method and the basis sets are very important to evaluate, not only the best level of theory but also the best basis set to be used to reproduce the experimental geometry and the vibrational frequencies. In previous studies on Cr compounds, such as oxotetrachlorochromate (V) anion [25], it was found that the inclusion of polarization functions in the basis sets significantly improved the theoretical geometry results and the lowest deviation, with reference to the experimental data, was obtained for the 6-31G* and 6-311G* basis sets and the B3PW91 functional [25]. In this case the lower difference between theoretical and experimental frequencies, measured by the root mean standard deviation (RMSD), was obtained with the combination B3LYP/6-31+G. In the study of the structures and vibrational spectra of chromium oxo anions and oxyhalide compounds, Bell et al. [26] have found that the B3LYP/Lanl2DZ combination gives the best fit for the geometries and the observed vibrational spectra. However, in the chromyl nitrate case the results with the Lanl2DZ basis set were not satisfactory, for that reason here, that basis set was not considered. In this chapter, DFT calculations were used to study the structure and vibrational properties of the compound. The normal mode calculations were accomplished using a GVFF. The molecular force field for the chromyl acetate, considering the acetate group as monodentate ligand calculated by using the DFT/6-31G* and 6-311++G** combinations is well represented. The force field produces satisfactory agreement between the calculated and experimental vibrational frequencies of chromyl acetate. DFT normal mode assignments, in terms of the potential energy distribution, are in general accord with those obtained from the normal coordinate analysis.

Also, the nature of the two types of Cr–O and Cr ← O bonds of chromyl acetate in gas phase and in both CCl$_4$ and DMSO solutions, were systematically and quantitatively investigated by the NBO analysis [27–29]. In addition, the topological properties of electronic charge density are analyzed using Bader atoms in molecules theory (AIM) [30].

2.2 Structural Study

The molecular structure for monodentate and bidentate chromyl acetate are shown in Fig. 2.1 together with the numbering of the atoms. The initial B3LYP structures modeled with different basis sets for chromyl acetate were carried out using C$_2$ symmetries according to the experimental structure obtained by Marsden et al. [11] for chromyl nitrate by using diffraction data and HF calculations. Table 2.1 shows a comparison of the total energies and dipole moment values for chromyl acetate with the B3LYP method using different basis sets.

In all cases, the most structure stable is obtained using the B3LYP/6-311++G** method combined with a diffuse function basis set, while the structure with higher energy is obtained using B3LYP/STO-3G calculation.

The structures were compared with those reported in the literature for barium and sodium acetate [31].

All the calculations in gas phase were made using the GAUSSIAN 03 [32] set of programs. The starting point for the geometry optimization was modeled with the *GaussView* program [33]. Calculations were made using hybrid density functional methods. In the last technique, Becke's three parameter functional and non-local correlation provided by Lee–Yang–Parr's (B3LYP) [34, 35] expressions were used, as implemented in the GAUSSIAN programs. Different basis sets were used. For the compound we realized the normal mode analysis using 6-31G* and 6-311++G** basis sets.

The harmonic force field in Cartesian coordinates for chromyl acetate which resulted from the calculations were transformed into "natural" internal coordinates [36] by the MOLVIB program [37, 38]. The natural coordinates for monodentate and bidentate chromyl acetates have been defined as proposed by Fogarasi et al. [39]. The analysis as bidentate ligand was performed with the two basis sets considering the acetate groups as two rings of four members where the deformations and torsion coordinates of these groups have been defined as proposed by Fogarasi et al. [39].

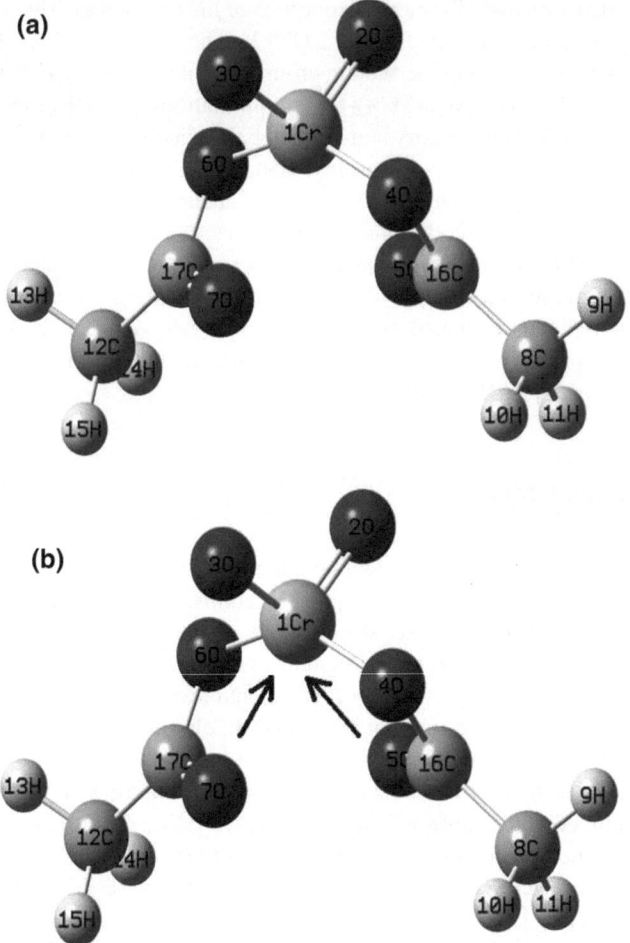

Fig. 2.1 The $C_2(1)$ molecular structure of chromyl acetate considering the acetate group as. **a** Monodentate ligand and **b** bidentate ligand

The numbering of the atoms for monodentate and bidentate chromyl acetates are described in Fig. 2.1

The scaled quantum mechanics (SQM) force field [40–42] was obtained using the transferable scale factors of Rauhut and Pulay [41] with the MOLVIB program [37, 38]. The potential energy distribution components (PED) larger than or equal to 10 % are subsequently calculated with the resulting SQM force field. The principal force constants for the compound were obtained.

An NBO analysis was then performed using the same basis sets with the NBO 3.1 program [43] included in GAUSSIAN 03 package programs [32]. The topological properties of the charge density in all systems studied were computed with the AIM2000 software [44].

Table 2.1 Total energy and dipole moment for chromyl acetate at B3LYP method

Basis set	ET (Hartree)	μ (D)
LANL2DZ	−693.68481616	3.09
STO-3G	−1633.03396705	2.47
3-21G*	−1643.67961761	3.27
6-31G	−1651.68610581	3.31
6-31G*	−1651.88702958	2.88
6-31+G	−1651.71932997	3.18
6-31+G*	−1651.91314474	2.81
6-311G*	−1652.10027672	2.54
6-311+G*	−1652.12267593	2.71
6-311++G**	−1652.13142696	2.73

The solvent effects were calculated with polarized continuum model (PCM) solvation model which were analyzed and compared with the obtained values in gas phase for the two studied basis sets. The use of an adequate basis set is important for the proper description of the solute–solvent interaction. Numerous authors have observed the dependency of the basis set on the SCRF calculated solvation [45, 46] and Pye et al. [47] have verified the greater changes induced adding diffuse function. Here, the B3LYP method using 6-31G* and 6-311++G** basis sets for SCRF calculations were used. To study the solvent effects the self-consistent reaction field method was employed. In this methodology, the solute molecule is located in a cavity surrounded by a continuum medium with a given dielectric constant. In this chapter, a value of 46.7 for the dielectric constant to simulate DMSO media and 2.228 for the dielectric constant to simulate CCl$_4$ media were used. In the simulation of the solvent effect, the self-consistent polarized continuum models have been used, as implemented in GAUSSIAN 03 and the cavity is created by a series of spheres, initially devised by Tomasi and Persico [48–50].

In addition, an NBO analysis for the two studied solvents were then performed with the NBO 3.1 program [43] using the 6-311++G** basis set and the topological properties of the charge density in two systems studied were computed with the AIM2000 software [44].

The dipole moment values do not follow a defined tendency as observed in Table 2.1. The calculation results with all basis sets used are given in Table 2.2.

The obtained theoretical parameters for the chromyl group were compared with the experimental values obtained for chromyl nitrate by Marsden et al. [11], while the calculated values for acetate group were compared with the recent experimental values obtained for acetate by Ibrahim et al. [51] and with those reported for sodium uranyl acetate [52], chromous acetate [53], sodium acetate [54], and anhydrous barium acetate [55]. In this way, the theoretical C–C and C–O bond distances are lower than the corresponding experimental values, while the C=O bond distance is longer than the observed value by Hsu et al. [54] for sodium acetate (1.245 Å).

Table 2.2 Comparison of calculated geometrical parameters at different levels of theory for chromyl acetate

B3LYP method

Atoms	STO-3G	3-21G*	6-31G	6-31G*	6-31+G	6-31+G*	6-311G*	6-311+G*	6-311++G**
Bond length (Å)									
(1,2)	1.484	1.563	1.576	1.551	1.579	1.557	1.549	1.554	1.554
(1,3)	1.484	1.563	1.576	1.551	1.579	1.557	1.549	1.554	1.554
(1,4)	1.859	1.875	1.925	1.913	1.920	1.913	1.909	1.916	1.917
(1,6)	1.859	1.875	1.925	1.913	1.920	1.913	1.909	1.916	1.917
(1,5)	2.169	2.330	2.323	2.261	2.343	2.276	2.286	2.293	2.292
(1,7)	2.169	2.330	2.323	2.261	2.343	2.276	2.286	2.293	2.292
(4,16)	1.356	1.351	1.342	1.30	1.344	1.310	1.306	1.307	1.307
(5,16)	1.302	1.259	1.269	1.245	1.268	1.246	1.238	1.239	1.239
(6,17)	1.356	1.351	1.342	1.309	1.344	1.310	1.306	1.307	1.307
(7,17)	1.302	1.259	1.269	1.245	1.268	1.246	1.238	1.239	1.239
(8,9)	1.102	1.094	1.094	1.095	1.090	1.095	1.092	1.092	1.092
(8,10)	1.102	1.095	1.096	1.095	1.090	1.096	1.093	1.092	1.093
(8,11)	1.100	1.089	1.090	1.091	1.091	1.091	1.088	1.088	1.088
(8,16)	1.544	1.490	1.485	1.497	1.486	1.497	1.495	1.494	1.494
(12,13)	1.102	1.094	1.094	1.095	1.095	1.095	1.092	1.092	1.092
(12,14)	1.102	1.095	1.096	1.095	1.096	1.096	1.093	1.093	1.093
(12,15)	1.100	1.089	1.090	1.091	1.091	1.091	1.088	1.088	1.088
(12,17)	1.544	1.490	1.486	1.497	1.486	1.497	1.495	1.494	1.494
Bond angle (°)									
(2,1,3)	107.5	107.1	108.1	107.9	108.3	108.2	107.9	108.2	108.2
(2,1,4)	101.9	103.5	103.7	104.0	103.9	104.3	103.8	104.2	104.2
(2,1,6)	98.1	99.9	98.6	97.9	98.7	97.9	98.3	98.2	98.1
(3,1,4)	98.1	99.9	98.6	979.208	98.7	97.9	98.3	98.2	98.1
(3,1,6)	101.9	103.5	103.7	104.0	103.9	104.3	103.8	104.2	104.2

(continued)

Table 2.2 (continued)

B3LYP method

Atoms	STO-3G	3-21G*	6-31G	6-31G*	6-31+G	6-31+G*	6-311G*	6-311+G*	6-311++G**
(4,1,6)	145.7	140.0	141.4	142.3	140.9	141.7	141.9	141.3	141.3
(1,4,16)	95.9	100.1	99.6	97.6	100.2	98.2	98.4	98.8	98.6
(1,6,17)	95.9	100.1	99.6	97.6	100.2	98.2	98.4	98.7	98.7
(9,8,10)	108.5	107.7	107.4	107.5	107.5	107.6	107.4	107.4	107.6
(9,8,11)	109.6	110.4	110.1	110.2	110.0	110.2	110.3	110.0	110.3
(9,8,16)	109.8	109.6	110.1	109.7	110.1	109.8	110.1	109.9	109.7
(10,8,11)	109.5	110.1	109.7	109.9	109.7	109.9	109.7	109.8	110.0
(10,8,16)	109.7	109.1	109.5	109.2	109.6	109.2	109.1	109.3	109.0
(11,8,16)	109.5	109.7	109.9	110.0	109.8	109.9	110.0	110.1	109.9
(13,12,14)	108.5	107.7	107.4	107.5	107.5	107.6	107.4	107.4	107.6
(13,12,15)	109.6	110.4	110.1	110.2	110.0	110.2	110.3	110.0	110.3
(13,12,17)	109.8	109.6	110.1	109.7	110.1	109.8	110.1	109.9	109.7
(14,12,15)	109.5	110.1	109.7	1.0	109.7	109.9	109.7	109.8	110.0
(14,12,17)	109.7	109.1	109.5	109.2	109.8	109.2	109.1	109.3	109.0
(15,12,17)	109.5	109.7	109.9	110.0	109.8	109.9	110.0	110.1	109.9
(4,16,5)	113.5	115.3	115.0	116.7	115.1	116.5	116.9	116.7	116.6
(4,16,8)	120.6	117.5	119.0	118.4	118.8	118.4	118.0	118.2	118.2
(5,16,8)	125.9	127.1	125.9	124.9	126.0	125.0	124.9	125.0	125.0
(6,17,7)	113.5	115.4	115.0	116.7	115.1	116.5	116.9	116.7	116.7
(6,17,12)	120.6	117.5	119.0	118.4.	118.8	118.4	118.0	118.2	118.2
(7,17,12)	125.9	127.1	125.9	124.9	126.0	125.0	124.9	125.0	125.0
Bond dihedral angle (°)									
(2,1,4,16)	−86.5	−86.2	−84.9	−83.8	−84.0	−83.3	−83.77	−83.5	−83.5
(3,1,4,16)	163.5	163.2	163.9	165.3	164.5	165.5	165.3	165.2	165.2
(6,1,4,16)	38.0	38.0	38.7	39.8	39.4	40.1	39.9	39.9	39.9

(continued)

Table 2.2 (continued)

B3LYP method

Atoms	STO-3G	3-21G*	6-31G	6-31G*	6-31+G	6-31+G*	6-311G*	6-311+G*	6-311++G**
(2,1,6,17)	163.5	163.2	163.9	165.3	164.5	165.5	165.3	165.2	165.2
(3,1,6,17)	-86.5	-86.2	-84.9	-83.8	-84.0	-83.3	-83.7	-83.5	-83.5
(4,1,6,17)	38.0	38.0	38.7	39.8	39.5	40.1	39.9	39.9	39.9
(1,4,16,5)	0.1	0.8	0.3	-0.5	-0.4	-0.6	-0.2	-0.4	-0.4
(1,4,16,8)	-178.9	-178.6	-179.0	-179.9	-179.9	179.8	-179.4	-180.0	-179.9
(1,6,17,7)	0.14	0.8	0.3	-0.5	-0.4	-0.6	-0.2	-0.4	-0.4
(1,6,17,12)	-178.9	-178.6	-179.0	-179.9	-179.9	179.8	-179.4	-180.0	-179.9
(9,8,16,4)	-58.3	-56.1	-54.6	-56.2	-55.5	-55.9	-52.6	-56.2	-55.6
(9,8,16,5)	122.7	124.4	126.2	124.3	124.9	124.4	128.2	124.2	124.8
(10,8,16,4)	60.8	61.6	63.3	615.294	62.5	17.713	65.0	61.5	62.0
(10,8,16,5)	-118.0	-117.8	-115.8	-117.9	-116.9	-117.8	-114.0	-118.0	-117.5
(11,8,16,4)	-178.8	-177.6	-176.1	-177.7	-176.9	-177.5	-174.5	-177.7	-177.2
(11,8,16,5)	22.6	29.5	46.9	28.4	35.9	29.3	63.6	27.1	32.6
(13,12,17,6)	-58.3	-56.1	-54.6	-56.1	-55.5	-55.9	-52.6	-56.2	-55.6
(13,12,17,7)	122.7	124.4	126.2	124.3	124.9	124.4	128.2	124.2	124.8
(14,12,17,6)	60.8	61.6	63.3	61.5	62.5	61.7	65.1	61.5	62.0
(14,12,17,7)	-118.0	-117.8	-115.8	-117.9	-116.9	-117.8	-114.0	-118.0	-117.5
(15,12,17,6)	-178.8	-177.6	-176.1	-177.6	-176.8	-177.5	-174.5	-177.7	-177.2
(15,12,17,7)	22.6	29.5	46.9	28.4	35.9	29.3	63.6	27.1	32.6

The calculated O–C–O bond angles with all the methods used are lower than the corresponding experimental values, while the O–C–C bond angles are longer than the reported values of 118 ° for sodium acetate [54]. According to these results, the method and basis set that best reproduces the experimental parameters for the chromyl acetate compound is B3LYP/6-31+G*, where the mean difference for bond lengths is 0.029 Å, while with B3LYP/3-21G* it is 4.1° for angles. The B3LYP functional gives somewhat less satisfactory agreement using the STO-3G (0.06 Å and 5.0°) and 6-31G (0.034 Å and 4.6°) basis sets. The inclusion of polarization functions, however, is important to have a better agreement with the experimental geometry: mean differences degrade to 0.031 Å and 4.34° for the 6-311++G** basis set. Similar to the experimental structure of chromyl nitrate [11], also with B3LYP calculations, we can represent the coordination around Cr as derived from a severely distorted octahedron where the acetate groups act as bidentate ligands and are asymmetrically bonded to Cr. The bond orders, expressed by Wiberg's indexes for chromyl acetate are given in Tables 2.3 and 2.4 and are compared with the values obtained for chromyl nitrate [10]. In this compound the chromium atom forms six bonds using 6-311++G** basis sets, two Cr=O bonds (bond order 1.9582), two Cr–O (bond order 0.5502), and two Cr ← O (bond order 0.1837). In chromyl nitrate the bonds order of these bonds are quite similar by using 6-311+G basis set, they are: two Cr=O (bond order 1.9721), two Cr–O (bond order 0.5374), and two Cr ← O (bond order 0.1821). The bond order value of this last bond was estimated by Marsden et al. [11] between 0.19 and 0.29, while for chromyl acetate it is slightly bigger than chromyl nitrate. Theoretically, in the B3LYP calculations the Cr–O–COO angle values are practically planar with the Cr atom with a variation of the dihedral angle between 0.1 and 0.8°, as can be seen in Table 2.2. In addition, in chromyl nitrate using B3LYP calculations and ab initio method, Marsden et al. [11] predict that the O=Cr=O bond angle in chromyl nitrate is smaller than the O–Cr–O angle, in contradiction with the valence-shell electron-pair (VSEPR) theory [56, 57]. In chromyl acetate a similar observation is found. This fact was explained on the basis of the delocalized and/or bonding characters of the relevant molecular orbitals (MO), as observed in the series of the $VO_2X_2^-$ anions [58] and chromyl nitrate [10]. The atomic orbital coefficients (AO) for Cr atom of chromyl acetate (d-type orbitals) using 6-31G * and 6-311++G** basis sets are observed in Table 2.5. For the chromyl acetate the strongest bonding MOs involving Cr d-type orbitals that seem to be sensitive to the geometry can be considered, in increasing energy, those numbered as 20 (HOMO-36), 21 (HOMO-36) and 22 (HOMO-37) calculated with 6-31G* basis set, while those numbered as 25 (HOMO-52), 25 (HOMO-54) and 25 (HOMO-54) calculated with 6-311++G** basis set tend to widen the O_4–Cr_1–O_8 angle (125.3° by using 6-31G* basis set) on a maximum overlapping basis.

Table 2.3 Wiberg bond index matrix of chromyl acetate at different levels of theory

Atoms		B3LYP method							
		6-31G* basis set							
		1	2	3	4	5	6	7	8	9
1.	Cr	0.0000	1.9419	1.9419	0.5180	0.1771	0.5180	0.1771	0.0219	0.0004
2.	O	1.9419	0.0000	0.2311	0.0728	0.0153	0.0650	0.0569	0.0038	0.0003
3.	O	1.9419	0.2311	0.0000	0.0650	0.0569	0.0728	0.0153	0.0017	0.0003
4.	O	0.5180	0.0728	0.0650	0.0000	0.1524	0.0247	0.0075	0.0208	0.0032
5.	O	0.1771	0.0153	0.0569	0.1524	0.0000	0.0075	0.0035	0.0364	0.0143
6.	O	0.5180	0.0650	0.0728	0.0247	0.0075	0.0000	0.1524	0.0005	0.0000
7.	O	0.1771	0.0569	0.0153	0.0075	0.0035	0.1524	0.0000	0.0003	0.0000
8.	C	0.0219	0.0038	0.0017	0.0208	0.0364	0.0005	0.0003	0.0000	0.9013
9.	H	0.0004	0.0003	0.0003	0.0032	0.0143	0.0000	0.0000	0.9013	0.0000
10.	H	0.0005	0.0003	0.0001	0.0035	0.0149	0.0001	0.0003	0.8976	0.0013
11.	H	0.0024	0.0002	0.0004	0.0073	0.0020	0.0002	0.0001	0.9129	0.0007
12.	C	0.0219	0.0017	0.0038	0.0005	0.0003	0.0208	0.0364	0.0000	0.0000
13.	H	0.0004	0.0003	0.0003	0.0000	0.0000	0.0032	0.0143	0.0000	0.0000
14.	H	0.0005	0.0001	0.0003	0.0001	0.0003	0.0035	0.0149	0.0000	0.0000
15.	H	0.0024	0.0004	0.0002	0.0002	0.0001	0.0073	0.0020	0.0000	0.0000
16.	C	0.0122	0.0078	0.0065	1.2133	1.5431	0.0014	0.0031	1.0192	0.0066
17.	C	0.0122	0.0065	0.0078	0.0014	0.0031	1.2133	1.5431	0.0002	0.0000

Atoms		10	11	12	13	14	15	16	17
1.	Cr	0.0005	0.0024	0.0219	0.0004	0.0005	0.0024	0.0122	0.0122
2.	O	0.0003	0.0002	0.0017	0.0003	0.0001	0.0004	0.0078	0.0065
3.	O	0.0001	0.0004	0.0038	0.0003	0.0003	0.0002	0.0065	0.0078
4.	O	0.0035	0.0073	0.0005	0.0000	0.0001	0.0002	1.2133	0.0014
5.	O	0.0149	0.0020	0.0003	0.0000	0.0003	0.0001	1.5431	0.0031
6.	O	0.0001	0.0002	0.0208	0.0032	0.0035	0.0073	0.0014	1.2133

(continued)

Table 2.3 (continued)

Atoms	10	11	12	13	14	15	16	17
7. O	0.0003	0.0001	0.0364	0.0143	0.0149	0.0020	0.0031	1.5431
8. C	0.8976	0.9129	0.0000	0.0000	0.0000	0.0000	1.0192	0.0002
9. H	0.0013	0.0007	0.0000	0.0000	0.0000	0.0000	0.0066	0.0000
10. H	0.0000	0.0008	0.0000	0.0000	0.0000	0.0000	0.0074	0.0001
11. H	0.0008	0.0000	0.0000	0.0000	0.0000	0.0000	0.0041	0.0000
12. C	0.0000	0.0000	0.9013	0.9013	0.8976	0.9129	0.0002	1.0192
13. H	0.0000	0.0000	0.8976	0.0000	0.0013	0.0007	0.0000	0.0066
14. H	0.0000	0.0000	0.9129	0.0013	0.0000	0.0008	0.0000	0.0074
15. H	0.0000	0.0000	0.9129	0.0007	0.0008	0.0000	0.0001	0.0041
16. C	0.0074	0.0041	0.0002	0.0000	0.0001	0.0000	0.0000	0.0007
17. C	0.0001	0.0000	1.0192	0.0066	0.0074	0.0041	0.0007	0.0000

B3LYP method

6-311++G** basis set

Atoms	1	2	3	4	5	6	7	8	9
1. Cr	0.0000	1.9582	1.9582	0.5502	0.1837	0.5502	0.1837	0.0246	0.0005
2. O	1.9582	0.0000	0.2815	0.0898	0.0157	0.0693	0.0704	0.0049	0.0004
3. O	1.9582	0.2815	0.0000	0.0693	0.0704	0.0898	0.0157	0.0030	0.0003
4. O	0.5502	0.0898	0.0693	0.0000	0.1536	0.0374	0.0083	0.0209	0.0035
5. O	0.1837	0.0157	0.0704	0.1536	0.0000	0.0083	0.0034	0.0368	0.0146
6. O	0.5502	0.0693	0.0898	0.0374	0.0083	0.0000	0.1536	0.0006	0.0000
7. O	0.1837	0.0704	0.0157	0.0083	0.0034	0.1536	0.0000	0.0005	0.0000
8. C	0.0246	0.0049	0.0030	0.0209	0.0368	0.0006	0.0005	0.0000	0.9166
9. H	0.0005	0.0004	0.0003	0.0035	0.0146	0.0000	0.0000	0.9166	0.0000
10. H	0.0008	0.0003	0.0002	0.0039	0.0152	0.0001	0.0003	0.9123	0.0000
11. H	0.0029	0.0003	0.0006	0.0069	0.0019	0.0002	0.0001	0.9296	0.0014
12. C	0.0246	0.0030	0.0049	0.0006	0.0005	0.0209	0.0368	0.0000	0.0007

(continued)

Table 2.3 (continued)

B3LYP method

6-311++G** basis set

Atoms	1	2	3	4	5	6	7	8	9
13. H	0.0005	0.0003	0.0004	0.0000	0.0000	0.0035	0.0146	0.0000	0.0000
14. H	0.0008	0.0002	0.0003	0.0001	0.0003	0.0039	0.0152	0.0000	0.0000
15. H	0.0029	0.0006	0.0003	0.0002	0.0001	0.0069	0.0019	0.0000	0.0000
16. C	0.0160	0.0124	0.0114	1.1993	1.5531	0.0022	0.0026	1.0162	0.0072
17. C	0.0160	0.0114	0.0124	0.0022	0.0026	1.1993	1.5531	0.0003	0.0000

Atoms	10	11	12	13	14	15	16	17
1. Cr	0.0008	0.0029	0.0246	0.0005	0.0008	0.0029	0.0160	0.0160
2. O	0.0003	0.0003	0.0030	0.0003	0.0002	0.0006	0.0124	0.0114
3. O	0.0002	0.0006	0.0049	0.0004	0.0003	0.0003	0.0114	0.0124
4. O	0.0039	0.0069	0.0006	0.0000	0.0001	0.0002	1.1993	0.0022
5. O	0.0152	0.0019	0.0005	0.0000	0.0003	0.0001	1.5531	0.0026
6. O	0.0001	0.0002	0.0209	0.0035	0.0039	0.0069	0.0022	1.1993
7. O	0.0003	0.0001	0.0368	0.0146	0.0152	0.0019	0.0026	1.5531
8. C	0.9123	0.9296	0.0000	0.0000	0.0000	0.0000	1.0162	0.0003
9. H	0.0014	0.0007	0.0000	0.0000	0.0000	0.0000	0.0072	0.0000
10. H	0.0000	0.0008	0.0000	0.0000	0.0000	0.0000	0.0082	0.0001
11. H	0.0008	0.0000	0.0000	0.0000	0.0000	0.0000	0.0040	0.0000
12. C	0.0000	0.0000	0.0000	0.9166	0.9123	0.9296	0.0003	1.0162
13. H	0.0000	0.0000	0.9166	0.0000	0.0014	0.0007	0.0000	0.0072
14. H	0.0000	0.0000	0.9123	0.0014	0.0000	0.0008	0.0001	0.0082
15. H	0.0000	0.0000	0.9296	0.0007	0.0008	0.0000	0.0000	0.0040
16. C	0.0082	0.0040	0.0003	0.0000	0.0001	0.0000	0.0000	0.0008
17. C	0.0001	0.0000	1.0162	0.0072	0.0082	0.0040	0.0008	0.0000

Table 2.4 Wiberg index and atomic charges of chromyl acetate at different levels of theory

Atoms	Numbers	6-31G*		6-311++G**	
		Atomic charges	Wiberg index	Atomic charges	Wiberg index
Cr	1	1.55826	5.3488	1.21848	5.4736
O	2	−0.34082	2.4044	−0.23001	2.5186
O	3	−0.34082	2.4044	−0.23001	2.5186
O	4	−0.64091	2.0905	−0.60230	2.1464
O	5	−0.65241	2.0272	−0.62687	2.0600
O	6	−0.64091	2.0905	−0.60230	2.1464
O	7	−0.65241	2.0272	−0.62687	2.0600
C	8	−0.77846	3.8167	−0.66706	3.8663
H	9	0.26943	0.9285	0.23782	0.9452
H	10	0.27229	0.9269	0.24133	0.9435
H	11	0.26468	0.9312	0.23204	0.9479
C	12	−0.77846	3.8167	−0.66706	3.8663
H	13	0.26943	0.9285	0.23782	0.9452
H	14	0.27229	0.9269	0.24133	0.9435
H	15	0.26468	0.9312	0.23204	0.9479
C	16	0.82706	3.8259	0.80581	3.8337
C	17	0.82706	3.8259	0.80581	3.8337

2.3 Calculations in Solution

Total (E) and solvation energies (ΔG) and dipole moment for chromyl acetate in CCl_4 and DMSO solutions are observed in Table 2.6 at B3LYP/6-31G* and B3LYP/6-311++G** levels of theory.

On the other hand, a comparison of the theoretical geometrical parameters in gas phase with the obtained values in solutions is observed in Table 2.7 The more stable structure in solution, the greater solvation energy values, and the lower dipole moment values using the two basis sets are obtained for the compound in DMSO solvent.

This behavior could be related probably to the partially ionic nature of the compound and then will explain the affinity of the solvent for the polar compound. In solution, the more important change in the geometrical parameters are observed in the distances where the Cr_1–O_5 and Cr_1–O_7 bond lengths are about 0.95 Å lower than the corresponding values in gas phase. Moreover, the variations are independent of the solvent because in the two cases considered the values are approximately the same. This observation is probably explained due to strong association of the molecules in solution media. Also, when the compound is solvated in the CCl_4 or DMSO solvents, the C–H bond lengths of the two acetate groups undergo a slight increase in approximately 0.4 Å and, as a consequence of the solvent effect, the C_8–C_{16} bond length decreases more strongly (0.402 Å) than the C_{12}–C_{17} distance. For this reason, the dihedral angles involved in the changes also undergo variations between 3.4 and 6.4°, as observed in Table 2.7.

Table 2.5 Atomic orbital coefficients (AO) for Cr atom of chromyl acetate at B3LYP level

6-31G*				6-311++G**			
N° orbital	Type orbital	HOMO-36	HOMO-37	N° orbital	Type orbital	HOMO-36	HOMO-37
18	6XX	−0.33178	0.00000	25	15D 0	0.15892	0.00000
19	6YY	0.07504	0.00000	26	15D+1	0.00000	0.21345
20	6ZZ	0.24774	0.00000	27	15D−1	0.00000	−0.11802
21	6XY	0.18994	0.00000	28	15D+2	−0.13523	0.00000
22	6XZ	0.00000	0.35911	29	15D−2	0.11457	0.00000
23	6YZ	0.00000	−0.19411	30	16D 0	0.16165	0.00000
24	7XX	−0.20429	0.00000	31	16D+1	0.00000	0.22423
25	7YY	0.04488	0.00000	32	16D−1	0.00000	−0.12192
26	7ZZ	0.16284	0.00000	33	16D+2	−0.14011	0.00000
27	7XY	0.11686	0.00000	34	16D−2	0.11908	0.00000
28	7XZ	0.00000	0.20458	35	17D 0	0.15837	0.00000
29	7YZ	0.00000	−0.11549	36	17D+1	0.00000	0.20227
30	8F0	0.01138	0.00000	37	17D−1	0.00000	−0.11466
31	8F+1	0.00000	0.00858	38	17D+2	−0.12927	0.00000
32	8F−1	0.00000	−0.00403	39	17D−2	0.10925	0.00000
33	8F+2	0.00416	0.00000	40	18S	−0.00120	0.00000
34	8F−2	0.00292	0.00000	41	19PX	0.00000	0.02540
35	8F+3	0.00000	−0.00043	42	19PY	0.00000	0.04889
36	8F-3	0.00000	−0.00425	43	19PZ	0.06290	0.00000
				44	20PX	0.00000	−0.00099
				45	20PY	0.00000	0.00879
				46	20PZ	0.04372	0.00000
				47	21D 0	0.05593	0.00000
				48	21D+1	0.00000	0.00350
				49	21D-1	0.00000	−0.00900
				50	21D+2	0.01428	0.00000
				51	21D-2	−0.01112	0.00000

Table 2.6 Total (E) and solvation energies (ΔG) and dipole moment for chromyl acetate in different solvents

Solvents	B3LYP/PCM method					
	E (Hartree)		ΔG (Kcal/mol)		μ (D)	
	6-31G*	6-311++G**	6-31G*	6-311++G**	6-31G*	6-311++G**
DMSO	−1651.886	−1652.131	−6.93	−7.52	3.06	2.90
CCl₄	−1651.882	−1652.125	−2.83	−3.02	3.23	3.20

2.4 Topological Study

For chromyl acetate the intermolecular interactions have been analyzed using Bader's topological analysis of the charge electron density, $\rho(r)$ by means of the AIM program [44]. The localization of the critical points in the $\rho(r)$ and the values

Table 2.7 Comparison of theoretical geometrical parameters in gas phase and in CCl₄ and DMSO solutions at different levels of theory for chromyl acetate
[a]B3LYP method

Atoms	Gas Phase		CCl$_4$		DMSO	
	6-31G*	6-311++G**	6-31G*	6-311++G**	6-31G*	6-311++G**
Bond length (Å)						
(1,2)	1.551	1.554	1.553	1.555	1.554	1.555
(1,3)	1.551	1.554	1.553	1.555	1.554	1.555
(1,4)	1.913	1.917	1.912	1.916	1.912	1.915
(1,6)	1.913	1.917	1.912	1.916	1.912	1.915
(1,5)	2.261	2.292	1.310	1.308	1.311	1.308
(1,7)	2.261	2.292	1.247	1.241	1.248	1.240
(4,16)	1.300	1.307	1.310	1.308	1.311	1.308
(5,16)	1.245	1.239	1.247	1.241	1.248	1.240
(6,17)	1.309	1.307	1.095	1.092	1.095	1.092
(7,17)	1.245	1.239	1.095	1.093	1.096	1.093
(8,9)	1.095	1.092	1.091	1.088	1.091	1.088
(8,10)	1.095	1.093	1.495	1.491	1.492	1.491
(8,11)	1.091	1.088	1.095	1.092	1.095	1.092
(8,16)	1.497	1.494	1.095	1.093	1.096	1.093
(12,13)	1.095	1.092	1.091	1.088	1.091	1.088
(12,14)	1.095	1.093	1.495	1.491	1.492	1.491
(12,15)	1.091	1.088	1.553	1.555	1.554	1.555
(12,17)	1.497	1.494	1.553	1.555	1.554	1.555
Bond angle (°)						
(2,1,3)	107.9	108.2	108.0	108.3	108.1	108.2
(2,1,4)	104.0	104.2	104.2	104.5	104.6	104.4
(2,1,6)	97.9	98.1	97.6	97.9	97.2	97.9
(3,1,4)	97.9	98.1	97.6	97.9	97.2	97.9

(continued)

Table 2.7 (continued)

aB3LYP method

Atoms	Gas Phase		CCl$_4$		DMSO	
	6-31G*	6-311++G**	6-31G*	6-311++G**	6-31G*	6-311++G**
(3,1,6)	104.0	104.2	104.2	104.5	104.6	104.4
(4,1,6)	142.3	141.3	142.4	141.4	142.4	141.5
(1,4,16)	97.6	98.6	97.6	98.7	97.5	98.7
(1,6,17)	97.6	98.7	97.6	98.7	97.5	98.7
(9,8,10)	107.5	107.6	107.6	107.6	107.6	107.7
(9,8,11)	110.2	110.3	110.2	110.4	110.2	110.3
(9,8,16)	109.7	109.7	109.7	109.6	109.6	109.6
(10,8,11)	109.9	110.0	110.0	110.0	109.9	110.0
(10,8,16)	109.2	109.0	109.2	109.0	109.2	109.0
(11,8,16)	110.0	109.9	110.0	110.0	110.1	110.0
(13,12,14)	107.5	107.6	107.6	107.6	107.6	107.7
(13,12,15)	110.2	110.3	110.2	110.4	110.2	110.3
(13,12,17)	109.7	109.7	109.7	109.6	109.6	109.6
(14,12,15)	110.0	110.0	110.0	110.0	109.9	110.0
(14,12,17)	109.2	109.0	109.2	109.0	109.2	109.0
(15,12,17)	110.0	109.9	110.0	110.0	110.1	110.0
(4,16,5)	116.7	116.6	116.2	116.2	115.7	116.2
(4,16,8)	118.4	118.2	118.6	118.5	118.9	118.5
(5,16,8)	124.9	125.0	125.1	125.3	125.4	125.3
(6,17,7)	116.7	116.7	116.2	116.2	115.7	116.2
(6,17,12)	118.4	118.2	118.6	118.5	118.9	118.5
(7,17,12)	124.9	125.0	125.1	125.3	125.4	125.3

(continued)

Table 2.7 (continued)

[a]B3LYP method

Atoms	Gas Phase		CCl₄		DMSO	
	6-31G*	6-311++G**	6-31G*	6-311++G**	6-31G*	6-311++G**
Bond dihedral angle (°)						
(2,1,4,16)	−83.8	−83.5	−83.4	−83.1	−82.9	−83.0
(3,1,4,16)	165.3	165.2	166.0	166.0	166.2	165.7
(6,1,4,16)	39.8	39.9	40.2	40.3	40.6	40.4
(2,1,6,17)	165.3	165.2	166.0	166.0	166.2	165.7
(3,1,6,17)	−83.8	−83.5	−83.4	−83.1	−82.9	−83.1
(4,1,6,17)	39.8	39.9	40.2	40.3	40.6	40.4
(1,4,16,5)	−0.5	−0.4	−0.7	−0.6	−0.8	−0.7
(1,4,16,8)	−179.9	−179.9	179.7	179.8	179.6	179.6
(1,6,17,7)	−0.5	−0.4	−0.7	−0.6	−0.8	−0.7
(1,6,17,12)	−179.9	−179.9	179.7	179.8	179.6	179.6
(9,8,16,4)	−56.2	−55.6	−56.8	−56.1	−56.5	−56.4
(9,8,16,5)	124.3	124.8	123.6	124.3	124.0	123.9
(10,8,16,4)	61.5	62.0	60.9	61.5	61.1	61.2
(10,8,16,5)	−117.9	−117.5	−118.6	−118.0	−118.3	−118.4
(11,8,16,4)	−177.7	−177.2	−178.2	−177.7	−178.0	−177.9
(11,8,16,5)	28.4	32.6	22.0	27.3	24.9	24.2
(13,12,17,6)	−56.1	−55.6	−56.8	−56.1	−56.5	−56.4
(13,12,17,7)	124.3	124.8	123.6	124.3	124.0	123.9
(14,12,17,6)	61.5	62.0	60.9	61.5	61.1	61.2
(14,12,17,7)	−117.9	−117.5	−118.6	−118.0	−118.3	−118.4
(15,12,17,6)	−177.6	−177.2	−178.2	−177.7	−178.0	−177.9
(15,12,17,7)	28.4	32.6	22.0	27.3	25.0	24.2

[a] This work

Table 2.8 Analysis of bond critical points in chromyl acetate compared with chromyl nitrate

Chromyl nitrate

B3LYP/6-311++G

Bond	Cr_1-O_4	$Cr_1 \leftarrow O_6$	Cr_1-O_8	$Cr_1 \leftarrow O_{10}$	(3, +1)	(3, +1)
$\rho(r)$	0.379166	0.301963	0.379166	0.301963	0.0374596	0.0374596
$\nabla^2\rho(r)$	**−0.186240**	0.043910	**−0.186240**	0.043910	0.015960	0.015960
$\lambda 1$	−0.919694	−0.700007	−0.919694	−0.700007	−0.0342722	−0.0342722
$\lambda 2$	−0.852381	−0.641803	−0.852381	−0.641803	0.0578309	0.0578309
$\lambda 3$	1.585836	1.385788	1.585836	1.385788	0.1360750	0.1360750
$|\lambda 1|/\lambda 3$	0.57990	0.505200	0.57990	0.505200	0.250620	0.250620

Chromyl acetate

B3LYP/6−311++G**

Bond	Cr_1-O_4	$Cr_1 \leftarrow O_5$	Cr_1-O_6	$Cr_1 \leftarrow O_7$	(3, +1)	(3, +1)
$\rho(r)$	0.11650	0.04979	0.11650	0.04386	0.04029	0.04029
$\nabla^2\rho(r)$	**0.38672**	0.17678	**0.38357**	0.17678	0.18192	0.18180
$\lambda 1$	−0.22187	−0.05126	−0.22241	−0.05136	−0.04241	−0.04242
$\lambda 2$	−0.20548	−0.03874	−0.20727	−0.03891	0.05268	0.05275
$\lambda 3$	0.81409	0.26679	0.81326	0.26677	0.17164	0.17147
$|\lambda 1|/\lambda 3$	0.27254	0.19213	0.27348	0.19252	0.24708	0.24739

(Chromyl acetate in CCl$_4$) PCM/B3LYP/6-311++G**

Bond	Cr_1-O_4	$Cr_1 \leftarrow O_5$	Cr_1-O_6	$Cr_1 \leftarrow O_7$	(3, +1)	(3, +1)
$\rho(r)$	0.12971	0.039726	0.12971	0.03972	0.03776	0.03776
$\nabla^2\rho(r)$	**0.42561**	0.16205	**0.42240**	0.16183	0.16827	0.16814
$\lambda 1$	−0.25742	−0.04403	−0.25790	−0.04413	−0.03845	−0.03849
$\lambda 2$	−0.23598	−0.02945	−0.23778	−0.02956	0.03930	0.03934
$\lambda 3$	0.91902	0.23554	0.91809	0.23552	0.16744	0.16729
$|\lambda 1|/\lambda 3$	0.28010	0.18693	0.28090	0.18737	0.22963	0.23007

(Chromyl acetate in DMSO) PCM/B3LYP/6-311++G**

Bond	Cr_1-O_4	$Cr_1 \leftarrow O_5$	Cr_1-O_6	$Cr_1 \leftarrow O_7$	(3, +1)	(3, +1)
$\rho(r)$	0.11650	0.04386	0.11650	0.04386	0.04029	0.04029
$\nabla^2\rho(r)$	**0.38672**	0.17678	**0.38357**	0.17648	0.18192	0.18180
$\lambda 1$	−0.22187	−0.05126	−0.22241	−0.05136	−0.04241	−0.04242
$\lambda 2$	−0.20548	−0.03874	−0.20727	−0.03891	0.05268	0.05275
$\lambda 3$	0.81409	0.26679	0.81326	0.26677	0.17165	0.17148
$|\lambda 1|/\lambda 3$	0.27253	0.19213	0.27347	0.19252	0.24707	0.24737

The quantities are in atomic units

of the Laplacian at these points are important for the characterization of molecular electronic structure in terms of interactions nature and magnitude. In this study the results using 6-311++G** basis set were compared with those obtained for chromyl nitrate using 6-311++G basis set [10] and can be seen in Table 2.8 The analyses of the Cr ← O bonds' critical points for the compound studied are reported as well in gas phase as in solution calculations. In one case, the $Cr_1 \leftarrow O_5$ and $Cr_1 \leftarrow O_7$ bonds' critical points have the typical properties of the closed-shell interaction. That is, the values of $\rho(r)$ are relatively low (0.05 and 0.3 a.u.) as in chromyl nitrate while the $Cr_1 \leftarrow O_6$ and $Cr_1 \leftarrow O_{10}$ bonds' critical points, the relationship, $|\lambda 1|/\lambda 3$ are < 1 and the Laplacian of the electron density, $\nabla^2\rho(r)$ (0.04 and 0.2 a.u.), are positive indicating that the interaction is dominated by the contraction of charge away from the interatomic surface toward each nucleus [59–66]. The other important observation is related to the topological properties of the $Cr_1 \leftarrow O_4$ and $Cr_1 \leftarrow O_6$ bonds' critical points as shown in Table 2.8. In these cases, the electron density values are aproximately 0.1 a.u. while the positive values of the Laplacian of the electron density for the $Cr \leftarrow O$ bonds (0.4 a.u.), as observed in Table 2.8, indicate that the $Cr_1 \leftarrow O_4$ and $Cr_1 \leftarrow O_6$ bonds' critical points are found in a region of charge depletion. The interaction $Cr_1 \leftarrow O_4$ bond is the same as the $Cr_1 \leftarrow O_6$ bond, which has no characteristic of the shared inter-action, i.e., the value of electron density at the bond critical point is relatively high and the Laplacian of the charge density is positive indicating that the electronic charge is not concentrated in the internuclear region. These values of the Laplacian of the charge density do not agree with the values obtained for the $Cr_1 \leftarrow O_4$ and $Cr_1 \leftarrow O_8$ bonds in chromyl nitrate as can be seen in Table 2.8 [10]. The two last bonds in that compound are more covalently bonded than $Cr_1–O_4$ and $Cr_1–O_6$ in chromyl acetate. The different topological property values of the two $Cr \leftarrow O$ bonds for chromyl acetate are not in agreement with those above B3LYP level results analyzed for chromyl nitrate. Evidently the nature of both compounds is different between them.

The $\rho(r)$ and $\nabla^2\rho(r)$ at the critical points related to $Cr \leftarrow O$ bonds are not comparable with the respective 0.395 and 1.164 a.u. values reported for the $Cr–O$ bond critical point in the $CrOF_4$ compound [57]. The $\rho(r)$ and $\nabla^2\rho(r)$ at the critical points related to the $Cr_1–O_4$ and $Cr_1–O_6$ bonds are comparable with the respective 0.091 and 0.429 a.u. values reported for the $Cr–F_{eq}$ bond critical point in the $CrO_2F_4{}^{2-}$(cis) compound [57]. Moreover, the (3, +1) critical point as shown in Table 2.8 in chromyl acetate would confirm the nature of the two $Cr \leftarrow O$ bonds in the respective structure. The two ring points of the electron density obtained by AIM analysis reveals that the mode of coordination adopted for the acetate groups in chromyl acetate is monodentate. These results indicated that the character of $Cr \leftarrow O$ bonds are different in chromyl acetate from the nitrate compound in spite of the theoretical structure supporting the conclusions about the nature of the coordination of the Cr atom for this compound being similar to that observed by electron-diffraction experiments in gas phase in chromyl nitrate [10]. For the $Cr_1 \leftarrow O_5$ and $Cr_1 \leftarrow O_7$ bonds the values of $\rho(r) < 0.07$ a.u. are characteristic of ionic closed-shell interactions, while for the $Cr_1–O_4$ and $Cr_1–O_6$ bonds the large values of $\rho(r)$ are indicative of a strong shared interaction and, moreover, the slightly large positive values of $\nabla^2\rho(r)$ indicate a polar displacement toward oxygen consistent with the view that the MO bonds are strong polar simple bonds [57] different from the $Cr=O_2$ and $Cr=O_3$ bonds where large positive values of

$\rho(r)$ (0.303 a.u.) and $\nabla^2\rho(r)$ (0.996 a.u.) indicate a polar displacement toward oxygen consistent with strong polar double bonds [57].

2.5 Vibrational Study

The infrared spectra of the solid chromyl acetate in a KBr pellet and in d_6-DMSO solutions were taken from a previous study where the compound was obtained as reported in Ref. [24] and from our measurements [8]. The mentioned spectra were compared with the vibrational spectra of barium and sodium acetate obtained from the literature [31]. The structure for the compound has C_2 symmetry and 45 vibrational normal modes including 23 A + 22 B modes. All vibrational modes are infrared and Raman active. Table 2.9 shows the calculated harmonic frequencies for chromyl acetate using B3LYP method with different basis sets. Note that the lowest theoretical frequencies were not observed in the infrared spectrum and, for this reason, the lower frequencies of barium acetate were taken as experimental values. In all cases, the theoretical values were compared with the respective experimental values by means of the RMSD values. It can be seen that the best results are obtained with a B3LYP/6-311++G** calculation and that the introduction of diffuse functions (but not of polarization functions!) is essential to have a good approximation of the experimental values, especially in the case of the Cr=O and Cr–O stretchings. The calculated harmonic frequencies for chromyl acetate in CCl_4 and DMSO solvents using two basis sets are observed in Table 2.10. The important shiftings are observed in DMSO in relation to CCl_4 solvent and more specially related to bands associated with the observed change in the distances mentioned in the section corresponding to structural analysis. This analysis was performed taking into account both possibilities for acetate groups: monodentate and bidentate because it is impossible to differentiate between them on grounds of infrared and Raman spectra alone [7, 10, 17]. The observed frequencies and the assignment for chromyl acetate considering the coordination adopted by acetate groups as monodentate and bidentate are given in Table 2.11.

Vibrational assignments were made on the basis of the potential energy distributions (PED) in terms of symmetry coordinates and by comparison with molecules that contain similar groups [5, 7–17].

We will refer to the results obtained at B3LYP level with 6-31G* basis set because after scaling this method a satisfactory agreement is obtained between the calculated and the experimental vibrational frequencies of chromyl acetate. In general, the theoretical Infrared spectrum of chromyl acetate demonstrates good agreement with the experimental spectrum (see Fig. 2.2). Below, we discuss the assignment of the most important groups for the compounds studied considering the two coordination kinds.

Table 2.9 Calculated harmonic frequencies (cm^{-1}) for chromyl acetate using B3LYP method with different basis sets

STO-3G	3-21G*	6-31G	6-31+G	6-31+G*	6-31G*	6-311G*	6-311+G*	6-311++G**
3495	3183	3188	3181	3175	3182	3160	3158	3154
3495	3183	3188	3181	3175	3182	3160	3158	3154
3472	3130	3140	3135	3133	3139	3119	3117	3114
3472	3130	3140	3135	3133	3139	3119	3117	3114
3313	3074	3070	3066	3069	3074	3057	3056	3051
3313	3074	3070	3066	3069	3074	3057	3056	3051
1677	1609	1593	1577	1655	1670	1660	1647	1645
1676	1602	1585	1571	1650	1664	1654	1641	1639
1643	1534	1510	1506	1492	1497	1486	1485	1471
1643	1534	1510	1506	1492	1497	1486	1485	1470
1623	1516	1500	1494	1485	1491	1481	1479	1464
1622	1514	1499	1492	1484	1490	1480	1478	1463
1535	1451	1450	1445	1436	1444	1427	1423	1415
1535	1451	1450	1445	1434	1441	1425	1422	1413
1477	1313	1350	1328	1384	1392	1374	1368	1364
1464	1297	1338	1318	1377	1385	1366	1360	1356
1424	1158	1101	1099	1088	1110	1103	1084	1084
1415	1134	1101	1099	1086	1109	1102	1081	1078
1120	1107	1081	1067	1080	1092	1079	1078	1071
1120	1107	1065	1054	1078	1089	1079	1074	1070
1073	1036	1038	1034	1026	1029	1024	1022	1017
1073	1034	1036	1033	1026	1028	1024	1022	1016
952	907	929	923	959	963	961	959	958
949	900	923	917	955	958	956	954	954
741	727	706	707	719	720	724	719	719
715	726	701	702	713	713	716	713	713
579	602	600	607	614	611	617	619	617
555	602	597	603	611	608	614	615	614
555	545	519	518	523	523	528	522	522
537	531	506	506	512	513	519	512	512
508	449	434	426	446	454	457	445	446
414	385	382	379	382	384	387	381	381
379	325	319	315	323	327	326	320	320
333	287	280	279	282	284	290	281	281
317	280	263	258	262	266	268	260	260
267	230	223	218	229	233	235	227	228
254	208	202	191	202	210	203	199	199
252	169	168	158	182	193	181	176	177
240	158	163	158	176	181	178	172	172
164	141	141	136	148	148	150	146	146
115	109	101	98	102	105	104	100	100
109	92	77	74	76	87	80	74	74
84	76	57	56	72	78	76	63	63
57	69	57	56	50	53	56	46	46
56	48	33	−33	50	52	55	45	45

Table 2.10 Calculated harmonic frequencies (cm^{-1}) for chromyl acetate in CCl$_4$ and DMSO solvents using B3LYP method with different basis sets

Gas		DMSO		CCl$_4$	
6-31G*	6-311++G**	PCM 6-31G*	PCM 6-311++G**	PCM 6-31G*	PCM 6-311++G**
3182	3154	3178	3149	3191	3141
3182	3154	3178	3149	3191	3141
3139	3114	3131	3105	3138	3091
3139	3114	3131	3105	3138	3091
3074	3051	3067	3043	3074	3031
3074	3051	3067	3043	3074	3031
1670	1645	1658	1632	1603	1555
1664	1639	1652	1626	1597	1551
1497	1471	1497	1471	1493	1470
1497	1470	1497	1471	1493	1470
1491	1464	1491	1462	1485	1458
1490	1463	1490	1463	1484	1457
1444	1415	1444	1416	1421	1400
1441	1413	1442	1415	1421	1399
1392	1364	1389	1361	1299	1265
1385	1356	1382	1353	1288	1256
1110	1084	1101	1076	1083	1075
1109	1078	1100	1075	1082	1075
1092	1071	1084	1073	1076	1057
1089	1070	1083	1067	1073	1048
1029	1017	1028	1016	1009	997
1028	1016	1028	1016	1007	996
963	958	967	962	913	898
958	954	962	958	907	891

(continued)

Table 2.10 (continued)

Gas		DMSO		CCl$_4$	
6-31G*	6-311++G**	PCM 6-31G*	PCM 6-311++G**	PCM 6-31G*	PCM 6-311++G**
720	719	726	725	731	731
713	713	717	718	725	727
611	617	617	624	631	643
608	614	614	620	628	640
523	522	526	525	540	543
513	512	513	513	533	537
454	446	455	446	454	449
384	381	388	384	391	391
327	320	330	322	322	323
284	281	284	281	299	298
266	260	267	261	280	276
233	228	237	231	241	235
210	199	209	199	235	237
193	177	199	182	189	192
181	172	187	179	184	183
148	146	151	150	161	165
105	100	115	113	129	154
87	74	103	101	92	151
78	63	101	84	89	132
53	46	81	77	69	92
52	45	68	68	68	90

Table 2.11 Experimental frequencies (cm^{-1}) for chromyl acetate

Experimental[a]		Assignment[c]				
IR solid	IR solution	Chromyl acetate[a]	SQM[b]	Monodentate	SQM[b]	Bidentate
2965		$v CH_3$	2997	v_a CH_3 ip	3057	v_a CH_3
			2997	v_a CH_3op	3057	v_a CH_3
2947			2956	v_a CH_3 ip	3016	v_a CH_3
			2956	v_a CH_3op	3016	v_a CH_3
2930		$v CH_3$	2894	v_s CH_3 ip	2953	v_s CH_3
			2894	v_s CH_3 op	2953	v_s CH_3
1714	1660	v_a C=O (monodentate)	1624	v_s C=O i p	1604	v_a COO ip
1610	1621	v_a C=O (bridge)	1612	v_a C=O op	1600	v_a COO op
1540	1575		1499	δa CH_3 ip	1430	δ CH_3
			1499	δa CH_3 op	1430	δ CH_3
1453	1453	v_a C=O (bridge)	1492	δa CH_3 ip	1425	δ CH_3
			1490	δa CH_3 op	1424	δ CH_3
	1432		1422	δs CH_3 i p	1385	$v_s COO$ ip
			1421	δs CH_3 op	1380	δs CH_3
1417	1371	v_s C=O (monodentate)	1378	v_s C-O	1333	δs CH_3
1352		δs CH_3	1366	v_a C-O	1326	$v_s COO$ op
			1098	ρCH_3 ip	1125	v_s Cr=O
			1097	ρCH_3 op	1122	v_a Cr=O
1045		ρCH_3	1024	ρCH_3 ip	1041	ρCH_3
			1024	ρCH_3 op	1041	ρCH_3
	977	$v Cr=O$	1003	v_s Cr=O	986	ρCH_3
	945	$v Cr=O$	995	v_a Cr=O	986	ρCH_3
900		v C-C	953	v_s Cr=O, δs O-Cr-O	933	v C-C
883 sh		v C-C	948	v_a Cr=O, δa O-Cr-O	930	δCOOip
755 br			723	δCOOip, vC-C	714	v C-C
668	677	δCOO	714	δCOOop, vC-C	704	δCOOop
655 sh			643	γCOOip	586	γCOOip
	628		642	γCOOop	583	γCOOop
619	618	ρ C-COO in-plane	508	ρCOOip	519	ρCOOip
	609		506	ρCOOop	506	ρCOOop
440	440	ρ C-COO out-plane	481	δCrO_2	438	δCrO_2
410	407		391	Wag CrO_2	376	v_a Cr-O
			326	v_a Cr-O	323	v_s Cr-O
			304	v_s Cr-O	260	δa O-Cr-O
			296	δa O-Cr-O	245	ρ CrO_2
			256	δs O-Cr-O	214	$\tau w CrO_2$
			217	δs Cr-O-C	195	$v Cr-O$
			214	τCOOip	189	τ C-O
			212	δa Cr-O-C	176	$v Cr-O$

(continued)

Table 2.11 (continued)

Experimental[a]		Assignment[c]				
IR solid	IR solution	Chromyl acetate[a]	SQM[b]	Monodentate	SQM[b]	Bidentate
			152	δs O–Cr–O	137	δs O–Cr–O
			108	τCOOop	97	v C ← O
			88	τC–O	81	τCOOop
			78	τC–O	71	v C ← O
			60	τwCH$_3$	48	τwCH$_3$
			56	τwCH$_3$	48	τwCH$_3$

Abbreviations v stretching, δ deformation, ρ rocking; *wag*, (γ) wagging, τw torsion, *a* antisymmetric, *s* symmetric, *op* out of phase, *ip* in phase
[a] Ref [8]
[b] From B3LYP/6-31G*
[c] This work

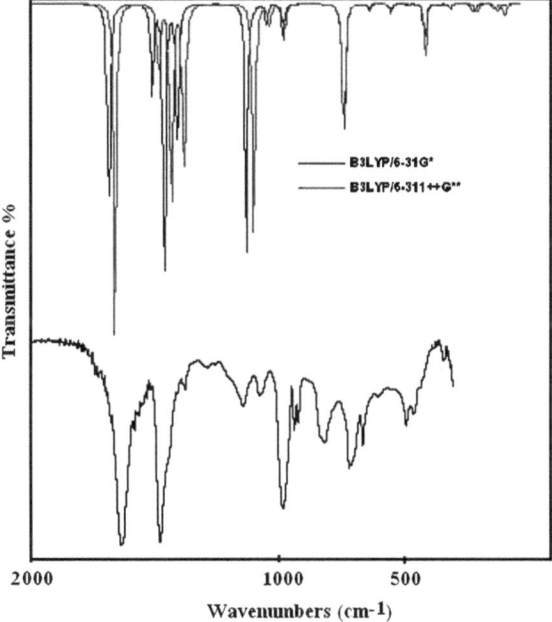

Fig. 2.2 Theoretical Infrared spectrum of CrO$_2$(CH$_3$COOH)$_2$ at B3LYP/6-31G* and B3LYP/6-311++G** theory levels (*upper*) compared with the experimental in solid phase (*bottom*)

2.6 Coordination Monodentate of the Acetate Groups

Practically all the vibration modes considering monodentate coordination of the acetate groups for chromyl acetate are perfectly characterized by the DFT/B3LYP/ 6-31G* and B3LYP/6-311++G** calculations and the contribution PED values almost do not change with the used method. The definition of natural internal coordinates for chromyl acetate with monodentate coordination adopted for acetate groups appear in Table 2.12.

2.7 CH$_3$ Modes

The IR bands at 2,965, 2,947, and 2,930 cm^{-1} are associated with the CH$_3$ stretchings and are calculated in all cases as pure modes. The in-phase modes have A symmetries while the out-of-phase modes have B symmetries. In the infrared spectrum of acetic acid these modes are observed at 3,051, 2,996, and 2,944 cm^{-1} [51], while in the infrared spectrum of barium acetate these modes appear at 3,128, 2,974, and 2,937 cm^{-1} [31]. The frequencies corresponding to CH$_3$ bending, rocking, and twisting modes are observed at expected frequencies. In the acetic acid, the CH$_3$ bending modes are calculated using 6-31G* basis set at 1,490, 1,484, and 1,423 cm^{-1} and observed at 1,430 and 1,382 cm^{-1} [51]. In chromyl acetate the PED values show that these modes are calculated as pure modes and for this reason they are assigned to the bands at 1,453 and 1,432 cm^{-1}. In acetic acid, the rocking modes are predicted, using a 6-31G* basis set, at 1,070 and 1,002 cm^{-1} and are observed at 1,048 and 989 cm^{-1} [51]. In chromyl acetate, these modes are calculated using 6-31G* basis set at 1,098, 1,097, 1,029, and 1,024 cm^{-1}. The modes at 1,098 and 1,029 cm^{-1} are assigned to CH$_3$ rocking modes in-phase while the remains are the corresponding out-of-phase modes. The CH$_3$ twisting modes in chromyl acetate are calculated at 60 and 56 cm^{-1} and the PED value indicate that these modes are strongly coupled with other modes but in this region with higher PED %. In acetic acid the CH$_3$ twisting mode is calculated using a 6-31G* basis set at 79 cm^{-1} and observed at 75 cm^{-1} [51].

2.8 Carboxylate Modes

The C = O antisymmetric and symmetric stretching modes reported in a previous paper [8] of CrO$_2$(CH$_3$COO)$_2$ solid were assigned to the strong bands observed in the infrared spectrum at 1,714 and 1,610 cm^{-1}, respectively. In acetic acid, this mode was calculated at 1,855 and observed at 1,788 cm^{-1} [51]. In this case, the two C = O stretching modes are calculated by the B3LYP/6-31G* method at 1,670 and 1,664 cm^{-1}. The frequencies predicted for these vibrational modes

Table 2.12 Definition of natural internal coordinates for chromyl acetate with monodentate coordination adopted for acetate groups

Symmetry A

S_1 = 2s (12-15) − s (12-14) − s (12-13) + 2s (8-11) − s (8-9) − s (8-10)   va (CH$_3$) ip

S_2 = s (12-13) − s (12-14) + s (8-9) − s (8-10)   va (CH$_3$) ip

S_3 = s (12-15) + s (12-14) + s (12-13) + s (8-11) + s (8-9) + s (8-10)   vs (CH$_3$) ip

S_4 = q (16-5) + q (17-7)   vs (C=O)

S_5 = β (11-8-9) − β (11-8-10) + β (13-12-15) − β (14-12-15)   δa (CH$_3$) ip

S_6 = 2 β (9-8-10) − β (11-8-9) − β (11-8-10) + 2 β (14-12-13) − β (13-12-15) − β (14-12-15)   δa (CH$_3$) ip

S_7 = q (16-4) + q (17-6)   v (C–O) ip

S_8 = β (9-8-10) + β (11-8-9) + β (11-8-10) − α (16-8-9) -α (16-8-10) − α (16-8-11) + β (14-12-13) + β (13-12-15) + β (14-12-15) − α (17-12-14) − α (17-12-15)   δs (CH$_3$) ip

S_9 = q (1-2) + q (1-3)   vs (Cr=O)

S_{10} = α (16-8-9) − α (16-8-10) + α (17-12-13) − α (17-12-14)   ρ (CH$_3$) ip

S_{11} = 2α (16-8-11) − α (16-8-9) -α (16-8-10) + 2 α (17-12-15) − α (17-12-13) − α (17-12-14)   ρ (CH$_3$) ip

S_{12} = r (12-17) + r (8-16)   v (C–C) ip

S_{13} = 2ψ (4-16-5) − ψ (4-16-8) − ψ (5-16-8) − ψ (6-17-7) − ψ (6-17-12) − ψ (7-17-12)   δ (O=C–O) ip

S_{14} = γ (8-16-5-4) + γ (12-17-7-6)   γ (COO) ip

S_{15} = t (1-4) + q (1-6)   vs (Cr–O)

S_{16} = θ (2-1-3)   δ (CrO$_2$)

S_{17} = ψ (4-16-8) − ψ (5-16-8) − ψ (7-17-12) + ψ (6-17-12)   ρ (COO) ip

S_{18} = φ (3-1-4) − φ (3-1-6) + φ (2-1-6) − φ (2-1-4)   ρ (CrO$_2$)

S_{19} = τ (3-1-4-16) + τ (2-1-4-16) + τ (2-1-6-17) + τ (3-1-6-17)   τ (C–O) ip

S_{20} = θ (1-4-16) + θ (1-6-17)   δ (Cr–O–C) ip

S_{21} = φ (6-1-4)   δ (O–Cr–O)

S_{22} = τ (1-4-16-8) + τ (1-4-16-5) − τ (1-6-17-12) − τ (1-6-17-7)   τ (COO) op

S_{23} = τ (4-16-8-9) + τ (4-16-8-10) + τ (4-16-8-11) + τ (6-17-12-13) + τ (6-17-12-14) + τ (6-17-12-15)   τ (CH$_3$) ip

Symmetry B

S_{24} = 2s (12-15) − s (12-14) − s (12-13) − 2s (8-11) + s (8-9) + s (8-10)   va (CH$_3$) op

S_{25} = s (12-14) − s (12-13) + s (8-9) − s (8-10)   va (CH$_3$) op

(continued)

Table 2.12 (continued)

S_{26} = s (12-15) + s (12-14) + s (12-13) − s (8-11) − s (8-9) − s (8-10)	vs (CH₃) op
S_{27} = q (16-5) − q (17-7)	va (C=O) op
S_{28} = β (11-8-9) − β (11-8-10) − β (13-12-15) + β (14-12-15)	δa (CH₃) op
S_{29} = 2β (9-8-10) − β (11-8-9) − β (11-8-10) − 2 β (14-12-13) + β (13-12-15) + β (14-12-15)	δa (CH₃) op
S_{30} = q (16-4) − q (17-6)	$ν$ (C-O) op
S_{31} = β (9-8-10) + β (11-8-9) + β (11-8-10) − α (16-8-9) -α (16-8-10) − α (16-8-11) − β (14-12-13) − β (13-12-15) − β (14-12-15) + α (17-12-13) + α (17-12-14) + α (17-12-15)	δs (CH₃) op
S_{32} = q (1-2) − q (1-3)	va (Cr=O)
S_{33} = α (16-8-10) − α (16-8-9) + α (17-12-13) − α (17-12-14)	ρ (CH₃) op
S_{34} = 2α (16-8-11) − α (16-8-9) -α (16-8-10) − 2 α (17-12-15) + α (17-12-13) + α (17-12-14)	ρ (CH₃) op
S_{35} = r (12-17) − r (8-16)	$ν$ (C-C) op
S_{36} = 2ψ (4-16-5) − ψ (4-16-8) − ψ (5-16-8) − 2 ψ (6-17-7) + ψ (6-17-12) + ψ (7-17-12)	δ (O=C-O) op
S_{37} = γ (8-16-5-4) − γ (12-17-7-6)	γ (COO) op
S_{38} = t (1-4) − q (1-6)	va (Cr-O)
S_{39} = ψ (4-16-8) − ψ (5-16-8) + ψ (7-17-12) − ψ (6-17-12)	ρ (COO) op
S_{40} = φ (3-1-4) + φ (2-1-4) − φ (3-1-6) − φ (2-1-6)	Wag (CrO₂)
S_{41} = φ (3-1-4) − φ (2-1-4) + φ (3-1-6) − φ (2-1-6)	τ (CrO₂)
S_{42} = θ (1-4-16) − θ (1-6-17)	δ (Cr-O-C) op
S_{43} = τ (1-4-16-8) + τ (1-4-16-5) − τ (1-6-17-12) − τ (1-6-17-7)	τ (COO) ip
S_{44} = τ (3-1-4-16) + τ (2-1-4-16) − τ (2-1-6-17) − τ (3-1-6-17)	τ (C-O) ip
S_{45} = τ (4-16-8-9) + τ (4-16-8-11) + τ (6-17-12-13) + τ (6-17-12-14) + τ (6-17-12-15)	τ (CH₃) op

Abbreviations q Cr=O bond distance, t Cr-O bond distance, s C–H bond distance, $θ$ = O=Cr=O bond distance, $φ$ = O–Cr–O bond angle, $ψ$ = C–C–O bond angles, $β$ = H–C–H bond angle, $α$, C–C–H bond angle.

Abbreviations $ν$ stretching, $δ$ deformation, $ρ$ in the plane bending or rocking, $γ$ out of plane bending or wagging, w twisting, a antisymmetric, s symmetric, ip in phase, op out of phase

show that the two stretching modes are split by about 6 cm^{-1} with the 6-31G* basis set. Both B3LYP methods underestimate the C = O stretching frequencies as compared to the experimental values. The modes are calculated as pure modes and both modes are assigned to the bands at 1,714 and 1,610 cm^{-1}, respectively, as in a previous paper [8]; moreover, the difference among them is of 9 cm^{-1}. Normally, the antisymmetric C = O stretching frequencies of the metal acetates are observed between 1,610 and 1,540 cm^{-1}, while the corresponding symmetric C = O stretching frequencies are observed among 1,451 and 1,394 cm^{-1} [12, 15, 19–22]. In the NaCH$_3$COO compound [12] the antisymmetric C = O stretching is observed at 1,556 cm^{-1}, in Ni(CH$_3$COO)$_2$ compound at 1,600 cm^{-1} [12], while this vibration mode in Cu(CH$_3$COO)$_2$ [22] is observed at 1,610 cm^{-1}. Agambar et al. [22] have assigned the δ(COO) mode for NaCH$_3$COO compound to the IR band at 645 cm^{-1}, while Spinner et al. [17] for the same compound have assigned the two δ(COO) in-phase modes to the Raman bands at 652 and 622 cm^{-1}, respectively. The calculations for acetic acid show the δ(COO) and wagging (COO) modes at 682 and 582 cm^{-1}, respectively, and were experimentally assigned to the bands at 657 and 642 cm^{-1}, respectively [51]. The theoretical calculations for chromyl acetate predict the two δ(COO) in-phase and out-of-phase modes at 723 and 714 cm^{-1}, respectively, whereas the corresponding wagging (COO) modes are calculated at 643 and 642 cm^{-1}. All the modes appear coupled with other modes. In chromyl acetate, the two δ(COO) modes are assigned to the bands at 755 and 678 cm^{-1}, respectively, while the two wagging (COO) modes are assigned to the shoulder and IR band at 655 and 628 cm^{-1}, respectively. The ρ(COO) in-phase and out-of-phase modes for chromyl acetate could be assigned to the bands at 619 and 609 cm^{-1}. Both modes have different symmetries and appear slightly coupled with other modes as the PED values predict. In acetic acid compound the twisting (COO) mode is calculated at 419 cm^{-1} and experimentally observed at 565 cm^{-1} [51], while this mode for NaCH$_3$COO was assigned by Spinner et al. [17] at 247 cm^{-1}. In a previous paper of chromyl acetate this mode was not assigned but, in this case the theoretical calculations predict the two twisting (COO) in-phase and out-of-phase modes at 214 and 108 cm^{-1}, respectively. In this low region in the Raman spectrum of numerous compounds that contain carboxylate groups, such NaCH$_3$COO, Ba(CH$_3$COO)$_2$ [31], and benzoic acid compounds [67, 68], it is possible to observe many bands with higher intensity probably due to additional resonance interactions. In the benzoic acid, Florio et al. [67] have observed them as a consequence of the fact that normal mode OH bend vibrations of cyclic dimer have contributions from both the OH and CH bends. This leads to a sharing of intensities over many states, which fills in the lower frequency region of the spectrum for benzoic acid dimer. Another cause would be as observed by Reva et al. [68] in the same cyclic dimer that the other more strongly H-bonded dimer may have more substantial contributions from the coupling of the OH stretch with the intermolecular stretch.

2.9 Chromyl Modes

In chromyl nitrate the unscaled DFT frequencies for the symmetric Cr=O stretchings mode are higher than the frequencies of the antisymmetric Cr=O stretchings [9, 10]. In this compound these modes are uncoupled with other modes. In other chromyl compounds these modes appear in 1,050–900 cm^{-1} region, i.e., in $CrO_2(ClO_4)_2$ they appear at 990 and 980 cm^{-1} [68], in $CrO_2(SO_3F)_2$ appear at 1,061 and 1,020 cm^{-1} [70], and in CrO_2F_2 and CrO_2Cl_2 they are observed for the first compound at 1,016 and 1,006 cm^{-1} while for the second one at 1,002 and 995 cm^{-1} respectively [70, 71]. The predicted frequencies for the vibrational modes of chromyl acetate show that the antisymmetric and symmetric Cr=O stretchings are slightly split indicating a small contribution of the central Cr atom in these vibrations. In this compound, the symmetric Cr=O stretching mode was also calculated at higher frequency (1,003 cm^{-1}) than the corresponding antisymmetric mode (995 cm^{-1}). The intensities of these bands using 6-31G* and 6-311++G** basis sets are not predicted correctly because the more intense band is related to the antisymmetric mode as it was observed in chromyl nitrate [10]. Previously, the two modes were assigned to the band observed in the infrared spectrum of the solid compound at 948 cm^{-1} [9]. In this case both modes are assigned to the IR bands observed in the spectrum of the sample in solution at 977 and 945 cm^{-1}.

The theoretical antisymmetric and symmetric Cr–O stretching modes in CrO_2 $(ClO_4)_2$ are observed, respectively, at 380 and 355 cm^{-1} [68] and in $CrO_2(NO_3)_2$ they were assigned these modes at 460 and 446 cm^{-1}, respectively [9, 10]. In this case the theoretical calculation predicts the antisymmetric and symmetric Cr–O stretching modes at 326 and 304 cm^{-1}, in form similar to that observed in chromyl nitrate [10].

The CrO_2 bending mode is observed in CrO_2F_2 at 364 cm^{-1} while in CrO_2Cl_2 it is observed at 356 cm^{-1} [71, 72]. In this chapter, the B3LYP/6-31G* method calculates the CrO_2 bending at 481 cm^{-1} in form similar to chromyl nitrate (453 cm^{-1}) [10]. With the other basis set this mode appears also coupled. Here, the IR band at 440 cm^{-1} was assigned to the CrO_2 bending. The other O–Cr–O bending modes are calculated at 296 and 256 cm^{-1} with 6-31G* basis set.

The wagging, rocking, and twisting modes of the CrO_2 group are not assigned in a previous paper [9]. In this case the calculations predict these modes in the low frequencies region and all modes are coupled with other modes of the acetate groups. The wagging CrO_2 mode is calculated using 6-31G* basis set at higher frequency (391 cm^{-1}) and with higher contribution PED than the rocking mode (296 cm^{-1}).

The CrO_2 twisting mode was not assigned previously. The PED values indicate that this mode is strongly coupled with vibrations of the same group and the acetate group. In this case this mode could be assigned at 256 cm^{-1} because it appears with a higher PED value.

2.10 Coordination Bidentate of the Acetate Groups

2.10.1 Acetate Groups as Rings of Four Members

In this case we performed calculations with the two basis sets considering both acetate groups as a ring of four members where the deformations and torsion coordinates of these groups have been defined as proposed by Fogarasi et al. [39] and are observed in Table 2.13. In this case a notable change in the assignment, in relation to the above monodentate considerations, is observed.

2.10.2 CH₃ Modes

In this case, a CH_3 bending mode is predicted at 1,333 cm^{-1}, while two CH_3 rocking modes are calculated at 986 cm^{-1} different from the monodentate coordination, as observed in Table 2.11 The twisting modes are observed at expected frequencies and calculated at 48 cm^{-1}.

2.10.3 Acetate Group

For this group, the more significant difference in relation to the monodentate case is presented. In this case, a C–O symmetric stretching in-phase mode is clearly predicted at 1,385 cm^{-1}, while one of the two $\delta(COO)$ modes is calculated at 930 cm^{-1} in reference to the above monodentate type, as shown in Table 2.11 The expected C ← O antisymmetric and symmetric modes are calculated at 97 and 71 cm^{-1}, while one of the two twisting (COO) modes is clearly calculated at 81 cm^{-1}.

2.10.4 Chromyl Group

For this group the calculations predict the two Cr=O stretchings at frequencies different from the monodentate case, the antisymmetric Cr=O stretchings are calculated at 1,122 cm^{-1}, while the corresponding symmetric stretchings are at 1,125 cm^{-1}.

The CrO_2 bending and the remains modes of this group are observed at practically the same frequencies as the above monodentate coordination.

For this analysis we think that the monodentate coordination is possible for this compound because the two Cr=O stretching modes should be observed with

Table 2.13 Definition of natural internal coordinates for chromyl acetate with bidentate coordination adopted for acetate groups

Symmetry A

$S_1 = 2s\,(12\text{-}15) - s\,(12\text{-}14) - s\,(12\text{-}13) + 2s\,(8\text{-}11) - s\,(8\text{-}9) - s\,(8\text{-}10)$	$va\,(CH_3)$ ip
$S_2 = s\,(12\text{-}13) - s\,(12\text{-}14) + s\,(8\text{-}9) - s\,(8\text{-}10)$	$va\,(CH_3)$ ip
$S_3 = s\,(12\text{-}15) + s\,(12\text{-}14) + s\,(12\text{-}13) + s\,(8\text{-}11) + s\,(8\text{-}9) + s\,(8\text{-}10)$	$vs\,(CH_3)$ ip
$S_4 = q\,(16\text{-}5) - q\,(16\text{-}4) + q\,(17\text{-}6) - q\,(17\text{-}7)$	$va\,(C\text{-}O)$ ip
$S_5 = \beta\,(11\text{-}8\text{-}9) - \beta\,(11\text{-}8\text{-}10) + \beta\,(13\text{-}12\text{-}15) - \beta\,(14\text{-}12\text{-}15)$	$\delta a\,(CH_3)$ ip
$S_6 = 2\beta\,(9\text{-}8\text{-}10) - \beta\,(11\text{-}8\text{-}9) - \beta\,(11\text{-}8\text{-}10) + 2\,\beta\,(14\text{-}12\text{-}13) - \beta\,(13\text{-}12\text{-}15) - \beta\,(14\text{-}12\text{-}15)$	$\delta a\,(CH_3)$ ip
$S_7 = q\,(16\text{-}5) + q\,(16\text{-}4) + q\,(17\text{-}6) + q\,(17\text{-}7)$	$vs\,(C\text{-}O)$ ip
$S_8 = \beta\,(9\text{-}8\text{-}10) + \beta\,(11\text{-}8\text{-}9) + \beta\,(11\text{-}8\text{-}10) - \alpha\,(16\text{-}8\text{-}9) - \alpha\,(16\text{-}8\text{-}10) - \alpha\,(16\text{-}8\text{-}11) + \beta\,(14\text{-}12\text{-}13)$ $+ \beta\,(13\text{-}12\text{-}15) + \beta\,(14\text{-}12\text{-}15) - \alpha\,(17\text{-}12\text{-}13) - \alpha\,(17\text{-}12\text{-}14) - \alpha\,(17\text{-}12\text{-}15)$	$\delta s\,(CH_3)$ ip
$S_9 = q\,(1\text{-}2) + q\,(1\text{-}3)$	$vs\,(Cr{=}O)$
$S_{10} = \alpha\,(16\text{-}8\text{-}9) - \alpha\,(16\text{-}8\text{-}10) + \alpha\,(17\text{-}12\text{-}13) - \alpha\,(17\text{-}12\text{-}14)$	$\rho\,(CH_3)$ ip
$S_{11} = 2\,\alpha\,(16\text{-}8\text{-}11) - \alpha\,(16\text{-}8\text{-}9) - \alpha\,(16\text{-}8\text{-}10) + 2\,\alpha\,(17\text{-}12\text{-}15) - \alpha\,(17\text{-}12\text{-}13) - \alpha\,(17\text{-}12\text{-}14)$	$\rho\,(CH_3)$ ip
$S_{12} = r\,(12\text{-}17) - r\,(8\text{-}16)$	$\nu\,(C\text{-}C)$ op
$S_{13} = r\,(12\text{-}17) + r\,(8\text{-}16)$	$\nu\,(C\text{-}C)$ ip
$S_{14} = \tau\,(5\text{-}1\text{-}4\text{-}16) + \tau\,(4\text{-}16\text{-}5\text{-}1) - \tau\,(1\text{-}4\text{-}16\text{-}5) - \tau\,(16\text{-}5\text{-}1\text{-}4) - \tau\,(7\text{-}1\text{-}6\text{-}17) + \tau\,(6\text{-}17\text{-}7\text{-}1)$ $- \tau\,(1\text{-}6\text{-}17\text{-}7) - \tau\,(17\text{-}7\text{-}1\text{-}6)$	$\tau\,(COO)$ ip
$S_{15} = \psi\,(4\text{-}16\text{-}8) - \psi\,(5\text{-}16\text{-}8) - \psi\,(7\text{-}17\text{-}12) + \psi\,(6\text{-}17\text{-}12)$	$\rho\,(COO)$ ip
$S_{16} = \phi\,(2\text{-}1\text{-}3)$	$\delta\,(CrO_2)$
$S_{17} = t\,(1\text{-}4) + t\,(1\text{-}6) + t\,(1\text{-}5) + t\,(1\text{-}7)$	$vs\,(Cr\text{-}O)$
$S_{18} = \phi\,(3\text{-}1\text{-}4) - \phi\,(3\text{-}1\text{-}6) + \phi\,(2\text{-}1\text{-}6) - \phi\,(2\text{-}1\text{-}4)$	$\rho\,(CrO_2)$
$S_{19} = t\,(1\text{-}4) + t\,(1\text{-}6) - t\,(1\text{-}5) - t\,(1\text{-}7)$	$va\,(Cr\text{-}O)$
$S_{20} = \tau\,(3\text{-}1\text{-}4\text{-}16) + \tau\,(2\text{-}1\text{-}4\text{-}16) + \tau\,(2\text{-}1\text{-}6\text{-}17) + \tau\,(3\text{-}1\text{-}6\text{-}17) + \tau\,(2\text{-}1\text{-}7\text{-}17) + \tau\,(3\text{-}1\text{-}4\text{-}17)$ $+ \tau\,(2\text{-}1\text{-}5\text{-}16) + + \tau\,(3\text{-}1\text{-}5\text{-}16)$	$\tau\,(C\text{-}C)$
$S_{21} = \phi\,(1\text{-}4\text{-}16) + \phi\,(1\text{-}6\text{-}17)$	$\delta\,(Cr\text{-}O\text{-}C)$ ip
$S_{22} = q\,(16\text{-}5) + q\,(16\text{-}4) - q\,(17\text{-}6) - q\,(17\text{-}5)$	$vs\,(C\text{-}O)$ op
$S_{23} = \tau\,(4\text{-}16\text{-}8\text{-}9) + \tau\,(4\text{-}16\text{-}8\text{-}10) + \tau\,(4\text{-}16\text{-}8\text{-}11) - \tau\,(6\text{-}17\text{-}12\text{-}13) - \tau\,(6\text{-}17\text{-}12\text{-}14) - \tau\,(6\text{-}17\text{-}12\text{-}15)$	$\tau\,(CH_3)$ op

(continued)

Table 2.13 (continued)

Symmetry B

$S_{24} = 2s\,(12\text{-}15) - s\,(12\text{-}14) - s\,(12\text{-}13) - 2s\,(8\text{-}11) + s\,(8\text{-}9) + s\,(8\text{-}10)$	$\nu a\,(CH_3)$ op
$S_{25} = s\,(12\text{-}14) - s\,(12\text{-}13) + s\,(8\text{-}9) - s\,(8\text{-}10)$	$\nu a\,(CH_3)$ op
$S_{26} = s\,(12\text{-}15) + s\,(12\text{-}14) + s\,(12\text{-}13) - s\,(8\text{-}11) - s\,(8\text{-}9) - s\,(8\text{-}10)$	$\nu s\,(CH_3)$ op
$S_{27} = q\,(16\text{-}4) - q\,(16\text{-}4) - q\,(17\text{-}6) + q\,(17\text{-}7)$	$\nu a\,(C\text{-}O)$ op
$S_{28} = \beta\,(11\text{-}8\text{-}9) - \beta\,(11\text{-}8\text{-}10) - \beta\,(13\text{-}12\text{-}15) + \beta\,(14\text{-}12\text{-}15)$	$\delta a\,(CH_3)$ op
$S_{29} = 2\,\beta\,(9\text{-}8\text{-}10) - \beta\,(11\text{-}8\text{-}9) - \beta\,(11\text{-}8\text{-}10) - 2\,\beta\,(14\text{-}12\text{-}13) + \beta\,(13\text{-}12\text{-}15) + \beta\,(14\text{-}12\text{-}15)$	$\delta a\,(CH_3)$ op
$S_{30} = \beta\,(9\text{-}8\text{-}10) + \beta\,(11\text{-}8\text{-}9) + \beta\,(11\text{-}8\text{-}10) - \alpha\,(16\text{-}8\text{-}9) - \alpha\,(16\text{-}8\text{-}10) - \alpha\,(16\text{-}8\text{-}11) - \beta\,(14\text{-}12\text{-}13)$ $- \beta\,(13\text{-}12\text{-}15) - \beta\,(14\text{-}12\text{-}15) + \alpha\,(17\text{-}12\text{-}13) + \alpha\,(17\text{-}12\text{-}15)$	$\delta s\,(CH_3)$ op
$S_{31} = q\,(16\text{-}5) + q\,(16\text{-}4) - q\,(17\text{-}6) - q\,(17\text{-}5)$	$\nu s\,(C\text{-}O)$ op
$S_{32} = q\,(1\text{-}2) - q\,(1\text{-}3)$	$\nu a\,(Cr\text{=}O)$
$S_{33} = \alpha\,(16\text{-}8\text{-}10) - \alpha\,(16\text{-}8\text{-}9) + \alpha\,(17\text{-}12\text{-}13) - \alpha\,(17\text{-}12\text{-}14)$	$\rho\,(CH_3)$ op
$S_{34} = 2\,\alpha\,(16\text{-}8\text{-}11) - \alpha\,(16\text{-}8\text{-}9) - \alpha\,(16\text{-}8\text{-}10) - 2\,\alpha\,(17\text{-}12\text{-}15) + \alpha\,(17\text{-}12\text{-}13) + \alpha\,(17\text{-}12\text{-}14)$	$\rho\,(CH_3)$ op
$S_{35} = \eta\,(4\text{-}1\text{-}5) + \eta\,(4\text{-}16\text{-}5) - \eta\,(1\text{-}7\text{-}17) - \eta\,(1\text{-}6\text{-}17) - \eta\,(1\text{-}4\text{-}16) - \eta\,(1\text{-}5\text{-}16) + \eta\,(6\text{-}1\text{-}7) + \eta\,(6\text{-}17\text{-}7)$	$\delta\,(O\text{=}C\text{-}O)$ ip
$S_{36} = \eta\,(4\text{-}1\text{-}5) + \eta\,(4\text{-}16\text{-}5) + \eta\,(1\text{-}7\text{-}17) + \eta\,(1\text{-}6\text{-}17) - \eta\,(1\text{-}4\text{-}16) - \eta\,(1\text{-}5\text{-}16) + \eta\,(6\text{-}1\text{-}7) + \eta\,(6\text{-}17\text{-}7)$	$\delta\,(O\text{=}C\text{-}O)$ op
$S_{37} = \gamma\,(8\text{-}16\text{-}5\text{-}4) - \gamma\,(12\text{-}17\text{-}7\text{-}6)$	$\gamma\,(COO)$ op
$S_{38} = \psi\,(4\text{-}16\text{-}8) - \psi\,(5\text{-}16\text{-}8) + \psi\,(7\text{-}17\text{-}12) - \psi\,(6\text{-}17\text{-}12)$	$\rho\,(COO)$ op
$S_{39} = t\,(1\text{-}6) + t\,(1\text{-}7) - t\,(1\text{-}4) - t\,(1\text{-}5)$	$\nu a\,(Cr\text{-}O)$
$S_{40} = \phi\,(2\text{-}1\text{-}7) + \phi\,(2\text{-}1\text{-}6) + \phi\,(3\text{-}1\text{-}7) + \phi\,(3\text{-}1\text{-}6) - \phi\,(2\text{-}1\text{-}4) - \phi\,(2\text{-}1\text{-}5)\text{-}\phi\,(3\text{-}1\text{-}4) - \phi\,(3\text{-}1\text{-}5)$	Wag (CrO_2)
$S_{41} = \phi\,(3\text{-}1\text{-}4) - \phi\,(2\text{-}1\text{-}5) + \phi\,(3\text{-}1\text{-}7) - \phi\,(2\text{-}1\text{-}6)$	$\tau\,(CrO_2)$
$S_{42} = t\,(1\text{-}6) - t\,(1\text{-}7) - t\,(1\text{-}4) + t\,(1\text{-}5)$	$\nu a\,(Cr\text{-}O)$
$S_{43} = \phi\,(7\text{-}1\text{-}4) - \phi\,(6\text{-}1\text{-}5)$	$\delta a\,(O\text{-}Cr\text{-}O)$
$S_{44} = \tau\,(5\text{-}1\text{-}4\text{-}16) + \tau\,(4\text{-}16\text{-}5\text{-}1) - \tau\,(1\text{-}4\text{-}16\text{-}5) - \tau\,(16\text{-}5\text{-}1\text{-}4) - \tau\,(7\text{-}1\text{-}6\text{-}17) - \tau\,(6\text{-}17\text{-}7\text{-}1) + \tau\,(1\text{-}6\text{-}17\text{-}7)$ $+ \tau\,(17\text{-}7\text{-}1\text{-}6)$	$\tau\,(COO)$ op
$S_{45} = \tau\,(4\text{-}16\text{-}8\text{-}9) + \tau\,(4\text{-}16\text{-}8\text{-}10) + \tau\,(4\text{-}16\text{-}8\text{-}11) + \tau\,(6\text{-}17\text{-}12\text{-}13) + \tau\,(6\text{-}17\text{-}12\text{-}14) + \tau\,(6\text{-}17\text{-}12\text{-}15)$	$\tau\,(CH_3)$ ip

*   as two rings of four members

$q = Cr\text{=}O$ bond distance, $r = Cr\text{-}O$ bond distance, $s = N\text{-}O$ bond distance, $\theta = O\text{=}Cr\text{=}O$ bond angle, $\phi = O\text{-}Cr\text{-}O$ bond angle, $\psi = O\text{=}Cr\text{-}O$ bond angles, $\beta = O\text{-}N\text{-}O$ bond angle

Abbreviations ν stretching, δ deformation, ρ in the plane bending or rocking, γ out of plane bending or wagging, τw twisting, a antisymmetric, s symmetric, *ip* in phase.*op* out of phase

Table 2.14 Comparison of scaled internal force constants for chromyl acetate

| | [A]Chromyl acetate | | | |
| | Monodentate | | [#]Bidentate | |
Coordinates	6-31G*	6-311++G**	6-31G*	6-311++G**
f (C = O)	10.38	10.38	12.19	11.94
f (C–O)	6.85	6.52	–	–
f (Cr = O)	9.09/(6.55)	8.63/(6.56)	9.09/(6.57)	8.63/(6.56)
f (Cr–O)	2.79/(6.09)	2.69/(7.34)	2.69/(1.38)	2.61/(1.27)
f (C–H)	4.99	4.92	4.99	4.92
f (C–C)	4.31	4.27	4.31	4.27
f (O = C–O)	1.66	1.57/0.93	2.12	–
f (O = Cr = O)	1.59/(2.53)	1.54/(2.26)	2.19/(1.62)	2.11/(1.63)
f (O–Cr–O)	1.00/(0.80)	0.94(0.74)	7.28/(0.65)	7.34/(0.66)
f (C–O–Cr)	1.85	1.51	–	–
f (H–C–H)	0.60	0.58	0.60	0.58
f (C = O)/(C–O)	1.76	1.81	4.67	4.62
f (C = O)/(Cr–O)	0.18	0.19	−1.78	−1.79
f (Cr–O)/(C–O)	0.84	0.78	−3.20	−3.22

Units are mdyn $\mathring{A}^{-1}$ for stretching and stretching–stretching interaction and mdyn $\mathring{A}$ rad^{-2} for angle deformations

[a] This work

Among parenthesis the force constants for chromyl nitrate [12]

higher intensity at 977 and 945 cm^{-1}. The two observed strong IR bands in that region justify the stretching modes.

2.11 Force Field

The SQM force field [35–37] was obtained using the transferable scale factors of Rauhut and Pulay [36] for the acetate group and with the MOLVIB program [32, 33]. For the chromyl group the scale factors used taken were of the chromyl nitrate [5]. The corresponding force constants were estimated using the scaling procedure of Pulay et al. [28], as mentioned before. The harmonic force fields in Cartesian coordinates were transformed into the local symmetry or "natural" coordinates proposed by Fogarasi et al. [31]. The calculated force constants for both coordination modes of the acetate groups are collected in Table 2.14.

The obtained force constant values for the two considered cases could indicate the presence of both coordination modes, but the biggest obtained value in the f (O–Cr–O) force constant would indicate that the existence of the coordination bidentate for chromyl acetate is impossible.

2.12 Conclusions

In this chapter an approximate normal coordinate analysis, considering the mode of coordination adopted by acetate groups as monodentate and bidentate, was proposed for chromyl acetate.

The assignments previously made [8] were corrected and completed in accordance with the present theoretical results. The assignments of the 45 normal modes of vibration corresponding to chromyl acetate are reported.

The NBO and AIM analyses show practically a monodentate coordination of the acetate groups in chromyl acetate and a probable ionic nature of the compound.

The 6-31G * and 6-311+G basis sets at the B3LYP level were employed to obtain the molecular force fields and the vibrational frequencies.

Acknowledgments This work was subsidized with grants from CIUNT (Consejo de Investigaciones, Universidad Nacional de Tucumán). The author thanks Prof. Tom Sundius for his permission to use MOLVIB.

References

1. M. Sowinska, A. Bartecki, Transition Met. Chem. **10**(2), 63–66 (1985)
2. M. Sowinska, J. Myrczek, A. Bartecki, Spectrosc. lett. **26**(7), 1295–1309 (1993)
3. F. Freeman, P.J. Cameron, R.H. DuBois, J. Org. Chem. **33**, 3970 (1968)
4. F. Freeman, R.H. DuBois, N.J. Yamachika, Tetrahedron **25**, 3441 (1969)
5. E.J. Parish, N. Aksara, T.L. Boos, E.S. Kaneshiro, J. Chem. Res. (S), 708–709 (1999)
6. J.R. Hanson, Nat. Prod. Rep. **18**, 282–290 (2001)
7. C. Szántay, Pure Appl. Chem. **62**(7), 1299–1302 (1990)
8. S.A. Brandán, A. Ben Altabef, E.L. Varetti, J. Argent. Chem. Soc. **87**(1/2), 89–96 (1999)
9. E.L. Varetti, S.A. Brandán, A. Ben Altabef, Vib. Spectrosc. **5**, 219 (1993)
10. S.A. Brandán, M.L. Roldán, C. Socolsky, A. Ben Altabef, Spectrochim. Acta, 66A (2007)
11. C.J. Marsden, K. Hedberg, M.M. Ludwig, G.L. Gard, Inorg. Chem. **30**, 4761 (1991)
12. D.A. Edwards, R.N. Hayward, Can. J. Chem. **46**, 3443–3446 (1968)
13. S.D. Robinson, M.F. Uttley, J. Chem. Soc. Dalton. **18**, 1912–1920 (1973)
14. A.R. Katritzky, J.M. Lagowski, J.A.T. Beard, Spectrochim. Acta **16**, 964–978 (1960)
15. G.B. Deacon, R.J. Phillips, Coord. Chem. Rev. **33**, 227–250 (1980)
16. F. Vratny, C.N.R. Rao, M. Dilling, Anal. Chem. **33**, 1455 (1961)
17. E. Spinner, J. Chem. Soc. **812**, 4217–4226 (1964)
18. K. Nakamoto, C. Udovich, J. Takemoto, J. Am. Chem. Soc. **92**, 3973 (1970)
19. C.D. Garner, R.G. Senior, T.J. King, J. Am. Chem. Soc. **98**(12), 3526–3529 (1970)
20. T.A. Stephenson, S.M. Morehouse, A.R. Powell, J.P. Heffer, G. Wilkinson, J. Chem. Soc. **667**, 3632–3640 (1965)
21. T.A. Stephenson, G. Wilkinson, J. Inorg. Nucl. Chem. **29**, 2122–2123 (1967)
22. C.A. Agambar, K.G. Orrell, J. Am. Chem. Soc. A, 897–904 (1969)
23. S.Z. Haider, M.H. Khungkar, K. De, J. Inorg. Nucl. Chem. **24**, 847–850 (1962)
24. H.L. Krauss, Angew. Chem. **70**, 502 (1958)
25. M.L. Roldán, S.A. Brandán, E.L. Varetti, A. Ben Altabef, Z. Anorg. Allg. Chem. **632**, 2495 (2006)
26. S. Bell, T.J. Dines, J. Phys. Chem. A **104**, 11403 (2000)
27. A.E. Reed, L.A. Curtiss, F. Weinhold, Chem. Rev. **88**, 899 (1988)
28. J.P. Foster, F. Weinhold, J. Am. Chem. Soc. **102**, 7211 (1980)

29. A.E. Reed, F. Weinhold, J. Chem. Phys. **83**, 1736 (1985)
30. R.F.W. Bader, *Atoms in Molecules. A Quantum Theory* (Oxford University Press, Oxford, 1990). ISBN 0198558651
31. Spectral Database for Organic Compounds SDBS, IR (S.Kinugasa, K.Tanabe and T.Tamura) Raman (K.Tanabe and J.Hiraishi)
32. Gaussian 03, Revision B.01, M. J. Frisch, J. A. Pople, Gaussian, Inc., Pittsburgh PA, 2003
33. A.B. Nielsen, A.J. Holder, *GaussView, User's Reference*, (Gaussian, Inc., Pittsburgh, 1997–1998)
34. A.D. Becke, J. Chem. Phys. **98**, 5648 (1993)
35. C. Lee, W. Yang, R.G. Parr, Phys. Rev., B **37**, 785 (1988)
36. P. Pulay, G. Fogarasi, F. Pang, J.E. Boggs, J. Am. Chem. Soc. **101**(10), 2550 (1979)
37. T. Sundius, J. Mol. Struct. **218**, 321 (1990)
38. T. Sundius, *MOLVIB: A Program for Harmonic Force Field Calculation*, QCPE Program No. 604, (1991)
39. G. Fogarasi, P. Pulay, In: Vibrational spectra and structure, Vol. 14, J. E. Durig (Ed.), (Elsevier, Amsterdam, 125) 1985
40. P. Pulay, G. Fogarasi, G. Pongor, J.E. Boggs, A. Vargha, J. Am. Chem. Soc. **105**, 7037 (1983)
41. G. Rauhut, P. Pulay, J. Phys. Chem. **99**, 3093 (1995)
42. G. Rauhut, P. Pulay, J. Phys. Chem. **99**, 14572 (1995)
43. E.D. Glendening, A.E. Reed, J.E. Carpenter, F. Weinhold, NBO Version 3.1
44. AIM 2000 designed by, University of Applied Sciences, Bielefeld, Germany
45. O. Tapia, J. Chim. Phys. **87**, 875 (1990)
46. S.A. Brandán, S.B. Díaz, R. Cobos Picot, E.A. Disalvo, A. Ben Altabef, Spectrochim. Acta. **66**, 1152–1164 (2007)
47. C.C. Pye, W.W. Rudolph, J. Phys. Chem. A **107**, 8746 (2003)
48. J. Tomasi, M. Persico, Chem. Rev. **94**, 2027 (1994)
49. J. Tomasi, in: C.J. Cramer, D.G. Truhlar (Eds.), (Am. Chem. Soc., Washington, 1994), 10
50. S. Miertus, E. Scrocco, J. Tomasi, Chem. Phys. **55**, 117 (1981)
51. M. Ibrahim, E. Koglin, Acta Chim. Slov. **51**, 453–460 (2004)
52. W.H. Zachariasen, H.A. Plettinger, Acta Cryst. **12**, 526–530 (1959)
53. J.N. Van Niekerk, F.R.L. Schoening, J.F. De Wet, Acta Cryst. **6**, 501–504 (1953)
54. L.Y. Hsu, C.E. Nordman, Acta Cryst. C **39**, 690–694 (1983)
55. I. Gautier-Luneau, A. Mosset, J. Solid State Chem. **73**(2), 473–479 (1988)
56. R.J. Gillespie (ed.), *Molecular Geometry* (Van Nostrand-Reinhold, London, 1972)
57. R.J. Gillespie, I. Bytheway, T.H. Tang, R.F.W. Bader, Inorg. Chem. **35**, 3954 (1996)
58. M. Fernández Gómez, A. Navarro, S.A. Brandán, C. Socolsky, A. Ben Altabef, E.L. Varetti, J. Mol. Struct. (THEOCHEM) **626**, 101 (2003)
59. S. Wojtulewski, S.J. Grabowski, J. Mol. Struct. **621**, 285 (2003)
60. S.J. Grabowski, Monat. für Chem. **133**, 1373 (2002)
61. R.F.W. Bader, J. Phys. Chem. A **102**, 7314 (1998)
62. P.L.A. Popelier, J. Phys. Chem. A **102**, 1873 (1998)
63. U. Koch, P.L.A. Popelier, J. Phys. Chem. **99**, 9747 (1995)
64. G.L. Sosa, N. Peruchena, R.H. Contreras, E.A. Castro, J. Mol. Struct. (THEOCHEM) **401**, 77 (1997)
65. G.L. Sosa, N. Peruchena, R.H. Contreras, E.A. Castro, J. Mol. Struct. (THEOCHEM) **577**, 219 (2002)
66. G. M. Florio, T.S. Zwier, E.M. Myshakin, K.D. Jordan, E.L. Sibert, J. Chem. Phys., **118**(4) (2003)
67. I.D. Reva, S.G. Stepanian, J. Mol. Struc. **349**, 337–340 (1995)
68. M. Chaabouni, T. Chausse, J.L. Pascal, J. Potier, J. Chem. Research **5**, 72 (1980)
69. S.D. Brown, G.L. Gard, Inorg. Chem. **12**, 483 (1973)
70. H. Siebert, *"Anwendungen der schwingungsspektroskopie in der Anorganische Chemie"*, (Springer, Berlin, 1966), 72
71. K. Nakamoto, Infrared and Raman Spectra of Inorganic and Coordination Compounds, 5° Ed., J. Wiley & Sons, Inc., New York, 1997

Chapter 3
Theoretical Study on the Structural and Vibrational Properties of Chromyl Perchlorate

Abstract In this chapter a structural and vibrational study for chromyl perchlorate was performed by using the available experimental infrared spectrum and theoretical calculations based on the density functional theory (DFT). The structural properties for the compound, such as the bonds order, charge-transfers, and topological properties were studied by means of the Natural Bond Orbital (NBO) and the Atoms in Molecules theory (AIM) investigation. Two stable structures of the compound were theoretically determined in the gas phase and probably these conformations are present in the solid phase. The harmonic vibrational wavenumbers for the optimized geometries were calculated at different theory levels. For a complete assignment of the compound infrared spectrum, the DFT calculations were combined with Pulay's scaled quantum mechanical force field (SQMFF) methodology in order to fit the theoretical wavenumbers, values to the experimental ones. The results were then used to predict the Raman spectra, for which there are no experimental data. An agreement between theoretical and available experimental results was found and a complete assignment of all the observed bands in the vibrational spectra was performed. The theoretical vibrational calculations allowed us to obtain a set of scaled force constants fitting the observed wavenumbers.

Keywords Chromyl perchlorate · Vibrational spectra · Molecular structure · Force field · DFT calculations

Adapted from Journal of Molecular Structure: THEOCHEM, 908/1-3, S. Brandan, Theoretical Study of the Structure and Vibrational Spectra of Chromyl Perchlorate, CrO2(ClO4)2,19-25, Copyright 2009, with permission from Elsevier

S. A. Brandán, *A Structural and Vibrational Investigation into Chromylazide, Acetate, Perchlorate, and Thiocyanate Compounds*, SpringerBriefs in Molecular Science, DOI: 10.1007/978-94-007-5754-7_3, © The Author(s) 2013

57

(a)   **(b)**

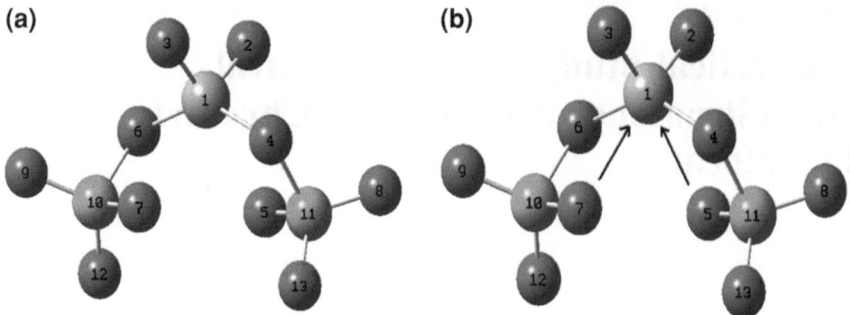

Fig. 3.1 The $C_2(1)$ molecular structure of chromyl perchlorate considering the perchlorate group as: **a** monodentate ligand and **b** bidentate ligand (Ref. [1]). reprinted from Journal of Molecular Structure: THEOCHEM, 908/1-3, S.A. Brandán, Theoretical Study of the Structure and Vibrational Spectra of Chromyl Perchlorate, $CrO_2(ClO_4)_2$, 19–25, copyright 2009, with permission from Elsevier

3.1 Introduction

The theoretical structure of chromyl perchlorate, $CrO_2(ClO_4)_2$ was recently studied using HF and DFT calculations [1, 2]. In this molecule the perchlorate ligands can act as monodentate and bidentate coordination as the nitrate and fluorosulfate groups act in nitrate and fluorosulfate chromyl compounds, respectively [3, 4].

The optimized geometries and frequencies for the normal modes of vibration were calculated using the HF, BLYP, B3LYP, B3P86, PBE1PBE, and B3PW91 methods. Two different C_2 structures for chromyl perchlorate were obtained; in one of them the perchlorate group could act as monodentate and bidentate ligands, and in the other only the perchlorate group could act as monodentate ligand. The normal mode calculations were accomplished using a generalized valence force field (GVFF) considering the two structures. Then a complete assignment for the compound was performed.

In this chapter, the principal results of the theoretical studies related to the structure and vibrational properties of chromyl perchlorate are presented.

3.2 Structural Analysis

Two different structures, both with C_2 symmetries named $C_2(1)$ and $C_2(2)$, were obtained from chromyl perchlorate by using the different methods considered with the 6-31G* basis set. In the $C_2(1)$ structure, the ClO_4^- groups can act as monodentate or bidentate ligands, which in the latter case, are asymmetrically bonded to Cr in a way similar to the experimental structure obtained for chromyl nitrate by Marsden et al. [5]. In the other structure, designed as $C_2(2)$, the perchlorate groups can act only as monodentate ligands. Both structures are observed in Figs. 3.1 and 3.2, respectively.

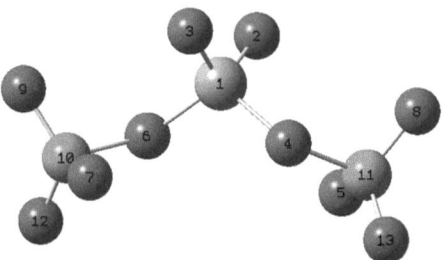

Fig. 3.2 The $C_2(2)$ molecular structure of chromyl perchlorate considering the perchlorate group as monodentate ligand (Ref. [1]). Reprinted from Journal of Molecular Structure: THEOCHEM, 908/1-3, S.A. Brandán, Theoretical Study of the Structure and Vibrational Spectra of Chromyl Perchlorate, $CrO_2(ClO_4)_2$, 19–25, copyright 2009, with permission from Elsevier

Table 3.1 shows a comparison of experimental data for chromyl nitrate with the calculated geometrical parameters for the $C_2(1)$ structure of chromyl perchlorate with 6-31G* basis set at different theory levels [1].

Table 3.2 shows the comparison of all calculated geometrical parameters at different theory levels for the $C_2(2)$ structure of chromyl perchlorate. According to these results, the methods that best reproduce the experimental geometrical parameters for chromyl perchlorate compound are B3P86 and PBE1PBE, where the mean difference for bond lengths are 0.058 and 0.057 Å, respectively, while with the PBEPBE method it is 4.35° for angles. The functional B3P86 shows a somewhat less satisfactory agreement (4.58°).

In the $C_2(1)$ structure of chromyl perchlorate, the perchlorate groups act as bidentate ligands and are asymmetrically bonded to Cr. By contrast, in the $C_2(2)$ structure the coordination around Cr of the perchlorate groups is only possible as monodentate ligands. Note that the $Cr_1–O_4–Cl_{11}$, $Cr_1–O_6–Cl_{10}$ and $O_4–Cr_1–O_6$ bond angles are different in both structures, as it is observed in Tables 3.1 and 3.2. In the $C_2(1)$ structure the $Cr_1–O_4–Cl_{11}$ and $Cr_1–O_6–Cl_{10}$ angle values vary approximately from 102 to 127°, being the bigger $O_4–Cr_1–O_6$ angle (from 141.6 to 143.8°) in relation to the other two. For the $C_2(2)$ structure, the situation is slightly different because the $O_4–Cr_1–O_6$ angle is lower enough (from 111.8 to 113°) than the other angles($Cr_1–O_4–Cl_{11}$ and $Cr_1–O_6–Cl_{10}$ angles) whose values are between 117 and 127°.

All calculations predict for both structures of chromyl perchlorate that the $O_4–Cr_1–O_6$ angles are higher than the $O_2=Cr_1=O_3$ bond angle in accordance with the results obtained by Marsden et al. [5] for chromyl nitrate. This contradiction with the valence-shell electron-pair repulsion (VSEPR) theory [6, 7] could be explained in a way similar to chromyl nitrate by means of Molecular Orbital (MO) studies by analyzing the delocalized and/or bonding characters of the relevant MO [1, 2, 8].

The NBO [9] and AIM studies [10–13] for the compound reveals that the coordination mode adopted for the perchlorate groups in the $C_2(1)$ structure of chromyl perchlorate is bidentate, while for the $C_2(2)$ structure with all methods used, the coordination of the perchlorate groups may represent only as monodentate ligands [1, 2].

Table 3.1 Comparison between the experimental and calculated geometrical parameters at different theory levels for chromyl perchlorate

	C$_2$(1) symmetry							Experimental
	Ab initio			DFT methods				
Parameters	HF	B3P86	B3PW91	PBEPBE	BLYP	PBE1PBE	SVWN5	Ref. [2]
Bond length (Å)								
(1,2)	1.478	1.537	1.538	1.567	1.476	1.531	1.550	1.586
(1,3)	1.478	1.537	1.538	1.567	1.476	1.531	1.550	1.586
(1,4)	1.909	1.887	1.8886	1.888	1.784	1.889	1.869	1.957
(1,6)	1.909	1.887	1.8886	1.888	1.784	1.889	1.869	1.957
(1,7)	2.298	2.309	2.333	2.398	1.596	2.303	2.400	2.254
(1,5)	2.298	2.309	2.333	2.398	1.419	2.303	2.400	2.254
(4,11)	1.545	1.623	1.6278	1.683	1.596	1.611	1.631	
(5,11)	1.467	1.500	1.4998	1.519	1.419	1.495	1.509	
(6,10)	1.545	1.623	1.6278	1.683	1.412	1.611	1.631	
(7,10)	1.467	1.500	1.4998	1.519	1.412	1.495	1.509	
(8,11)	1.409	1.446	1.4474	1.471	1.406	1.440	1.453	
(9,10)	1.409	1.446	1.4474	1.471	1.406	1.440	1.453	
(10,12)	1.405	1.444	1.4449	1.469	1.476	1.438	1.451	
(11,13)	1.405	1.444	1.4449	1.469	1.476	1.438	1.451	
Bond angle (°)								
(2,1,3)	105.2	106.8	106.8	107.5	107.9	106.7	107.4	112.6
(2,1,4)	104.6	105.3	105.3	106.1	110.8	105.2	106.0	104.5
(2,1,6)	97.1	96.3	96.5	96.4	107.6	96.5	95.2	97.2
(3,1,4)	97.1	96.3	96.5	96.4	107.6	96.5	95.2	105.4
(3,1,6)	104.6	105.3	105.3	106.1	110.8	105.2	106.0	104.5
(4,1,6)	143.8	143. 4	143.0	141.6	111.8	143.3	143.8	140.6
(1,4,11)	105.8	104.0	104.5	104.5	127.1	104.0	102.1	
(1,6,10)	105.8	104.0	104.5	104.5	127.1	104.0	102.1	
(6,10,7)	97.7	97.5	97.7	97.9	103.9	97.7	97.1	

(continued)

Table 3.1 (continued)

| Parameters | C$_2$(1) symmetry | | | | | | | Experimental |
| | Ab initio | DFT methods | | | | | | |
	HF	B3P86	B3PW91	PBEPBE	BLYP	PBE1PBE	SVWN5	Ref. [2]
(6,10,9)	108.6	107.9	107.8	107.4	104.7	107.9	108.0	
(6,10,12)	108.6	107.5	107.3	106.4	103.5	107.5	107.4	
(7,10,9)	112.5	113.3	113.4	113.9	113.6	113.1	113.4	
(7,10,12)	113.2	113.7	113.7	114.0	114.5	113.5	113.5	
(9,10,12)	114.6	115.0	114.9	115.1	114.7	115.0	115.3	
(4,11,5)	97.6	97.5	97.7	97.9	103.9	97.7	97.1	
(4,11,8)	108.6	107.9	107.9	107.4	104.7	107.9	108.0	
(4,11,13)	108.6	107.5	107.3	106.4	103.5	107.5	107.4	
(5,11,8)	112.5	113.3	113.4	113.9	113.6	113.2	113.4	
(5,11,13)	113.2	113.7	113.7	114.0	114.5	113.5	113.6	
(8,11,13)	114.6	115.0	114.9	115.1	114.7	115.0	115.3	
Dihedral angle (°)								
D(2,1,4,11)	−81.4	−80.0	−79.5	−78.3	−29.2	−80.5	−81.4	
D(3,1,4,11)	170.7	170.5	171.0	171.4	147.0	170.1	168.7	
D(6,1,4,11)	43.7	43.9	44.5	45.2	90.9	43.6	42.2	
D(2,1,6,10)	170.7	170.5	171.0	171.4	147.0	170.1	168.8	
D(3,1,6,10)	−81.4	−80.0	−79.5	−78.3	−29.2	−80.5	−81.4	
D(4,1,6,10)	43.7	43.9	44.5	45.1	90.9	43.6	42.3	
D(1,4,11,5)	−45.0	−52.8	−58.9	−67.9	−36.7	−48.3	−33.8	
D(1,4,11,8)	112.4	112.2	111.8	111.3	83.3	112.7	114.1	
D(1,4,11,13)	−122.2	−123.0	−123.8	−124.8	156.1	−122.6	−120.8	
D(1,6,10,7)	−45.0	−52.8	−58.9	−67.9	−36.1	−48.4	−33.8	
D(1,6,10,9)	112.4	112.3	111.8	111.3	83.4	112.7	114.1	
D(1,6,10,12)	−122.2	−123.0	−123.8	−124.8	156.1	−122.6	−120.9	

Table 3.2 Comparison between the calculated geometrical parameters at different theory levels for the chromyl perchlorate $C_2(2)$ symmetry

Methods	Ab initio		DFT					
Parameters	HF	B3P86	B3PW91	PBEPBE	BLYP	PBE1PBE	SVWN5	B3LYP
Bond length (Å)								
(1,2)	1.476	1.538	1.540	1.569	1.582	1.532	1.550	1.548
(1,3)	1.476	1.538	1.540	1.569	1.582	1532	1.550	1.548
(1,4)	1.784	1.770	1.773	1.781	1.786	1.772	1.755	1.779
(1,6)	1.784	1.770	1.773	1.781	1.786	1.772	1.755	1.779
(1,7)	1.596	1.712	1.716	1.786	3.385	1.694	1.728	1.740
(1,5)	1.419	1.453	1.454	1.479	3.385	1.448	1.462	1.464
(4,11)	1.596	1.712	1.716	1.786	1.844	1.694	1.728	1.740
(5,11)	1.419	1.453	1.454	1.479	1.496	1.448	1.462	1.464
(6,10)	1.412	1.449	1.450	1.475	1.844	1.443	1.456	1.459
(7,10)	1.412	1.449	1.450	1.475	1.496	1.443	1.456	1.459
(8,11)	1.406	1.446	1.447	1.472	1.493	1.440	1.454	1.456
(9,10)	1.406	1.446	1.447	1.472	1.493	1.440	1.454	1.456
(10,12)	1.476	1.538	1.540	1.569	1.490	1.532	1.550	1.548
(11,13)	1.476	1.538	1.540	1.569	1.490	1.532	1.550	1.548
Bond angle (°)								
(2,1,3)	107.9	109.6	109.6	110.5	110.9	109.4	110.6	109.8
(2,1,4)	110.8	109.7	109.8	109.7	109.7	109.8	109.1	109.9
(2,1,6)	107.6	107.6	107.6	107.2	107.1	107.5	107.4	107.6
(3,1,4)	107.6	107.6	107.6	107.2	107.1	107.5	107.4	107.6
(3,1,6)	110.8	109.7	109.8	109.7	109.7	109.8	109.1	109.9
(4,1,6)	111.8	112.5	112.3	112.5	112.3	112.6	113.3	112.0
(1,4,11)	127.1	123.2	123.9	120.9	121.7	122.8	117.2	123.5
(1,6,10)	127.1	123.2	123.9	120.9	121.7	122.8	117.1	123.5

(continued)

Table 3.2 (continued)

$C_2(2)$ symmetry

Methods	Ab initio		DFT					
Parameters	HF	B3P86	B3PW91	PBEPBE	BLYP	PBE1PBE	SVWN5	B3LYP
(6,10,7)	103.9	104.4	104.5	104.6	104.6	104.2	103.3	104.5
(6,10,9)	104.7	103.7	103.7	103.4	103.8	103.8	103.6	103.7
(6,10,12)	103.5	100.8	100.7	99.8	99.2	101.1	100.6	100.6
(7,10,9)	113.6	114.7	114.7	115.1	115.1	114.7	115.0	114.8
(7,10,12)	114.5	115.1	115.1	115.4	115.4	115.1	115.5	115.1
(9,10,12)	114.7	115.4	115.3	115.6	115.6	115.3	115.7	115.4
(4,11,5)	103.9	104.4	104.5	104.6	104.6	104.2	103.3	104.5
(4,11,8)	104.6	103.7	103.7	103.4	103.8	103.8	103.6	103.7
(4,11,13)	103.5	100.8	100.7	99.8	99.2	101.1	100.6	100.6
(5,11,8)	113.6	114.7	114.7	115.0	115.1	114.7	115.0	114.8
(5,11,13)	114.5	115.1	115.1	115.4	115.4	115.1	115.5	115.1
(8,11,13)	114.7	115.4	115.3	115.6	115.6	115.3	115.7	115.4
Dihedral angle (°)								
D(2,1,4,11)	−29.1	−27.5	−26.5	−30.1	−28.5	−29.0	−36.8	−26.6
D(3,1,4,11)	147.0	146.7	145.7	−150.1	−148.9	148.0	156.7	−146.2
D(6,1,4,11)	90.9	92.2	93.1	89.1	90.5	90.8	82.7	92.9
D(2,1,6,10)	147.0	146.7	145.7	−150.1	−148.9	148.0	156.7	−146.1
D(3,1,6,10)	−29.2	−27.5	−26.5	−30.1	−28.5	−29.0	−36.8	−26.6
D(4,1,6,10)	90.9	92.2	93.2	89.1	90.5	90.8	82.7	92.9
D(1,4,11,5)	−36.1	−38.8	−39.5	−38.8	−48.0	−37.9	−35.4	−40.2
D(1,4,11,8)	83.4	81.6	81.0	81.9	73.0	82.5	84.9	80.4
D(1,4,11,13)	156.1	158.6	159.3	−158.5	−167.5	157.6	155.0	−159.9
D(1,6,10,7)	−36.1	−38.8	−39.5	−38.8	−48.0	−37.9	−35.4	−40.2
D(1,6,10,9)	83.4	81.6	81.0	81.9	73.0	82.5	84.9	80.3
D(1,6,10,12)	156.1	158.6	159.3	−158.5	−167.5	−29.0	155.0	−159.9

The nature and the topological properties of both structures are different between them. In the $C_2(1)$ structure, the chromium atom forms six bonds, two Cr=O, two Cr–O and two Cr ← O, while in the $C_2(2)$ structure the Cr atom forms only four bonds.

3.3 Vibrational Analysis

The vibrational analysis for this compound was performed, for the $C_2(1)$ structure, by taking into account the monodentate and bidentate coordination modes [14–16, 17] for the perchlorate groups, while for the $C_2(2)$ structure only the coordination monodentate was considered. In this last case the perchlorate groups were considered with C_{3v} symmetry, while in the bidentate case the symmetry of both perchlorate groups is C_{2v}.

The definition of natural internal coordinates for the $C_2(1)$ structure of chromyl perchlorate with the monodentate and bidentate coordination modes adopted by perchlorate groups can be seen in Tables 3.3 and 3.4 respectively. Table 3.5 shows the definition of natural internal coordinates for the $C_2(2)$ structure of chromyl perchlorate only with the monodentate coordination.

Vibrational assignments were made on the basis of the potential energy distribution (PED) contributions in terms of symmetry coordinates and by comparison with molecules that contain similar groups [1, 2, 5, 18]. In all cases the theoretical values were compared to the respective experimental values by means of the RMSD values and the calculated frequencies not observed in the vibrational spectrum were taken as experimental values. The best results for the $C_2(2)$ structure are obtained with a B3LYP/6-31G* calculation, while for the $C_2(1)$ structure they are obtained at the B3P86/6-31G* level.

For the $C_2(1)$ structure, comparing both coordination modes, the SQM frequencies are slightly different for the monodentate and bidentate ligands. However, observing the PED values obtained with the B3P86/6-31G* method, a small difference between both coordination modes is found. Thus, for the monodentate coordination, there are 16 out of the 33 vibration modes whose PED contributions are higher than the other coordination mode. However, considering the bidentate coordination for the $C_2(1)$ structure, there are 17 out of the 33 vibration modes whose PED contributions are higher than the other coordination mode. By contrast, comparing the monodentate coordination for the $C_2(2)$ structure using the B3LYP/6-31G* calculation, only 25 out of the 33 expected modes have the PED contributions higher than the PED values for the $C_2(1)$ structure using B3P86/6-31G* method and for the same coordination mode.

For these observations, despite the infrared spectroscopy being limited to determine the mode of perchlorate coordination, it is possible by means of the GVFF study and observing the higher PED contribution to find the coordination mode of the perchlorate group. In this compound, probably both coordination modes are present in the solid phase due to two structures (bidentate for the

Table 3.3 Definition of natural internal coordinates for the $C_2(1)$ structure of chromyl perchlorate with the monodentate coordination adopted by perchlorate groups

A symmetry

$S_1 = 2s\ (9\text{-}10) + 2s\ (11\text{-}8) - s\ (10\text{-}7) - s\ (10\text{-}12) - s\ (11\text{-}5) - s\ (11\text{-}13)$ $\nu a\ (ClO_3)$ ip

$S_2 = s\ (11\text{-}13) + s\ (10\text{-}12) - s\ (11\text{-}5) - s\ (10\text{-}7)$ $\nu a\ (ClO_3)$ ip

$S_3 = q\ (1\text{-}2) - q\ (1\text{-}3)$ $\nu a\ (Cr{=}O)$

$S_4 = s\ (9\text{-}10) + s\ (10\text{-}12) + s\ (10\text{-}7) + s\ (11\text{-}5) + s\ (11\text{-}13) + s\ (11\text{-}8)$ $\nu s\ (ClO_3)$ ip

$S_5 = s\ (10\text{-}6) + s\ (11\text{-}4)$ $\nu s\ (Cl{-}O)$

$S_6 = r\ (1\text{-}4) + r\ (1\text{-}6)$ $\nu s\ (Cr{-}O)$

$S_7 = \alpha\ (12\text{-}10\text{-}7) + \alpha\ (9\text{-}10\text{-}12) + \alpha\ (9\text{-}10\text{-}7) - \alpha\ (13\text{-}11\text{-}5) - \alpha\ (8\text{-}11\text{-}5) - \alpha\ (8\text{-}11\text{-}13) - \beta\ (6\text{-}10\text{-}7)$
$-\beta\ (6\text{-}10\text{-}12) - \beta\ (9\text{-}10\text{-}6) + \beta\ (4\text{-}11\text{-}5) + \beta\ (8\text{-}11\text{-}4) + \beta\ (4\text{-}11\text{-}13)$ $\delta s\ (ClO_3)$ op

$S_8 = 2\alpha\ (12\text{-}10\text{-}7) - \alpha\ (9\text{-}10\text{-}12) - \alpha\ (9\text{-}10\text{-}7) + 2\alpha\ (13\text{-}11\text{-}5) - \alpha\ (8\text{-}11\text{-}5) - \alpha\ (8\text{-}11\text{-}13)$ $\delta a\ (ClO_3)$ ip

$S_9 = 2\alpha\ (9\text{-}10\text{-}12) - 2\alpha\ (13\text{-}11\text{-}5) + \alpha\ (8\text{-}11\text{-}13) - \alpha\ (9\text{-}10\text{-}7) + \alpha\ (8\text{-}11\text{-}5) - \alpha\ (9\text{-}10\text{-}12)$ $\delta a\ (ClO_3)$ op

$S_{10} = \theta\ (2\text{-}1\text{-}3)$ $\delta\ (O{=}Cr{=}O)$

$S_{11} = 2\beta\ (9\text{-}10\text{-}6) - 2\beta\ (8\text{-}11\text{-}4) - \beta\ (6\text{-}10\text{-}7) - \beta\ (13\text{-}11\text{-}4) - \beta\ (4\text{-}11\text{-}5) - \beta\ (6\text{-}10\text{-}12)$ $\rho\ (ClO_3)$ ip

$S_{12} = 2\beta\ (9\text{-}10\text{-}6) + 2\beta\ (8\text{-}11\text{-}4) - \beta\ (6\text{-}10\text{-}7) + \beta\ (13\text{-}11\text{-}4) + \beta\ (4\text{-}11\text{-}5) - \beta\ (6\text{-}10\text{-}12)$ $\rho\ (ClO_3)$ op

$S_{13} = \psi\ (3\text{-}1\text{-}4) + \psi\ (2\text{-}1\text{-}6) - \psi\ (3\text{-}1\text{-}6) - \psi\ (2\text{-}1\text{-}4)$ $\tau wis\ (CrO_2)$

$S_{14} = \phi\ (1\text{-}6\text{-}10) + \phi\ (1\text{-}4\text{-}11)$ $\delta s\ (Cr{-}O{-}Cl)$

$S_{15} = \rho\ (6\text{-}1\text{-}4)$ $\delta\ (O{-}Cr{-}O)$

$S_{16} = \tau\ (3\text{-}1\text{-}4\text{-}11) + \tau\ (2\text{-}1\text{-}4\text{-}11) + \tau\ (2\text{-}1\text{-}6\text{-}10) + \tau\ (3\text{-}1\text{-}6\text{-}10)$ $\tau\ (OCrOCl)$ ip

$S_{17} = \tau\ (1\text{-}4\text{-}11\text{-}5) + \tau\ (1\text{-}4\text{-}11\text{-}8) + \tau\ (1\text{-}4\text{-}11\text{-}13) + \tau\ (1\text{-}6\text{-}10\text{-}7) + \tau\ (1\text{-}6\text{-}10\text{-}9) + \tau\ (1\text{-}6\text{-}10\text{-}12)$ $\tau\ (ClO_3)$ ip

B symmetry

$S_{18} = 2s\ (9\text{-}10) - 2\ s\ (11\text{-}8) - s\ (10\text{-}7) - s\ (10\text{-}12) + s\ (11\text{-}5) + s\ (11\text{-}13)$ $\nu a\ (ClO_3)$ op

$S_{19} = s\ (10\text{-}12) + s\ (11\text{-}5) - s\ (10\text{-}7) - s\ (11\text{-}13)$ $\nu a\ (ClO_3)$ op

$S_{20} = q\ (1\text{-}2) + q\ (1\text{-}3)$ $\nu s\ (Cr{=}O)$

$S_{21} = s\ (9\text{-}10) + s\ (10\text{-}12) + s\ (10\text{-}7) - s\ (11\text{-}5) - s\ (11\text{-}13) - s\ (11\text{-}8)$ $\nu s\ (ClO_3)$ op

$S_{22} = s\ (10\text{-}6) - s\ (11\text{-}4)$ $\nu a\ (Cl{-}O)$

(continued)

Table 3.3 (continued)

S_{23} = r (1-4) − q (1-6)	va (Cr-O)
S_{24} = α (12-10-7) + α (9-10-12) + α (9-10-7) + α (13-11-5) + α (8-11-5) + α (8-11-13) − β (6-10-7) − β (6-10-12) − β (9-10-6) − β (4-11-5) − β (8-11-4) − β (4-11-13)	δs (ClO$_3$) ip
S_{25} = 2α (12-10-7) + 2α (13-11-5) − α (8-11-5) − α (8-11-13) − α (9-10-12) − α (9-10-7)	δa (ClO$_3$) ip
S_{26} = α (9-10-12) + α (8-11-5) − α (9-10-7) − α (8-11-13)	δa (ClO$_3$) op
S_{27} = β (6-10-7) + β (13-11-4) − β (4-11-5) − β (6-10-12)	ρ (ClO$_3$) ip
S_{28} = β (6-10-7) − β (13-11-4) + β (4-11-5) − β (6-10-12)	ρ (ClO$_3$) op
S_{29} = ψ (2-1-4) + ψ (3-1-4) − ψ (2-1-6) − ψ (3-1-6)	wag (CrO$_2$)
S_{30} = ψ (3-1-6) + ψ (3-1-4) − ψ (2-1-4) − ψ (2-1-6)	ρ (CrO$_2$)
S_{31} = φ (1-6-10) − φ (1-4-11)	δa (Cr-O-Cl)
S_{32} = τ (1-4-11-5) + τ (1-4-11-8) + τ (1-4-11-13) − τ (1-6-10-7) − τ (1-6-10-9) − τ (1-6-10-12)	τ (ClO$_3$) op
S_{33} = τ (3-1-4-11) + τ (2-1-4-11) − τ (2-1-6-10) − τ (3-1-6-10)	τ (OCrOCl) op

q = Cr=O bond distance, r = Cr–O bond distance, s = Cl–O bond distance, $θ$ = O=Cr=O bond angle, $ρ$ = O–Cr=O bond angle, $φ$ = Cr–O–Cl bond angle, $ψ$ = O=Cr–O bond angle, $α$ = O=Cr–O bond angles, $α$ = O=Cl=O bond angle, $β$ = O=Cl–O bond angle

Abbreviations v stretching, $δ$ deformation, $ρ$ in the plane bending or rocking, $γ$ out-of-plane bending or wagging, tw twisting, a antisymmetric, s symmetric, ip in phase, op out of phase

Table 3.4 Definition of natural internal coordinates for the $C_2(1)$ structure of chromyl perchlorate with the bidentate coordination adopted by perchlorate groups

A symmetry

$S_1 = s(9\text{-}10) + s(11\text{-}8) - s(10\text{-}12) - s(11\text{-}13)$ $va(Cl=O_2)$ ip

$S_2 = s(9\text{-}10) + s(10\text{-}12) + s(11\text{-}13) + s(11\text{-}8)$ $vs(Cl=O_2)$ ip

$S_3 = q(1\text{-}2) + q(1\text{-}3)$ $vs(Cr=O)$

$S_4 = s(10\text{-}6) - s(10\text{-}7) + s(11\text{-}4) - s(11\text{-}5)$ $va(Cl-O_2)$ ip

$S_5 = s(10\text{-}6) + s(10\text{-}7) + s(11\text{-}4) + s(11\text{-}5)$ $vs(Cl-O_2)$ ip

$S_6 = r(1\text{-}4) + r(1\text{-}5) + r(1\text{-}6) + r(1\text{-}7)$ $vs(Cr-O)$

$S_7 = 4\alpha(9\text{-}10\text{-}12) - 4\alpha(8\text{-}11\text{-}13) + \beta(8\text{-}11\text{-}4) + \beta(4\text{-}11\text{-}13) + \beta(8\text{-}11\text{-}5) + \beta(13\text{-}11\text{-}5)$

    $- \beta(12\text{-}10\text{-}7) - \beta(9\text{-}10\text{-}7) - \beta(6\text{-}10\text{-}12) - \beta(9\text{-}10\text{-}6)$ $\delta(Cl=O_2)$ op

$S_8 = \beta(7\text{-}10\text{-}12) + \beta(6\text{-}10\text{-}12) + \beta(13\text{-}11\text{-}5) + \beta(4\text{-}11\text{-}13) - \beta(9\text{-}10\text{-}6) - \beta(9\text{-}10\text{-}7)$

    $- \beta(8\text{-}11\text{-}4) - \beta(8\text{-}11\text{-}5)$ wag $(Cl=O_2)$ op

$S_9 = 4\alpha(9\text{-}10\text{-}12) + 4\alpha(8\text{-}11\text{-}13) - \beta(8\text{-}11\text{-}4) - \beta(4\text{-}11\text{-}13) - \beta(8\text{-}11\text{-}5) - \beta(13\text{-}11\text{-}5)$

    $- \beta(12\text{-}10\text{-}7) - \beta(9\text{-}10\text{-}7) - \beta(6\text{-}10\text{-}12) - \beta(9\text{-}10\text{-}6)$ $\delta(Cl=O_2)$ ip

$S_{10} = \theta(2\text{-}1\text{-}3)$ $\delta(O=Cr=O)$

$S_{11} = \beta(7\text{-}10\text{-}12) + \beta(9\text{-}10\text{-}6) + \beta(8\text{-}11\text{-}4) + \beta(13\text{-}11\text{-}5) - \beta(9\text{-}10\text{-}7) - \beta(6\text{-}10\text{-}12)$

    $- \beta(4\text{-}11\text{-}13) - \beta(8\text{-}11\text{-}5)$ $\tau(Cl=O_2)$ ip

$S_{12} = 4\alpha(6\text{-}10\text{-}7) + 4\alpha(5\text{-}11\text{-}4) - \beta(7\text{-}10\text{-}12) - \beta(9\text{-}10\text{-}7) - \beta(6\text{-}10\text{-}12) - \beta(9\text{-}10\text{-}6)$

    $- \beta(13\text{-}11\text{-}5) - \beta(4\text{-}11\text{-}13) - \beta(8\text{-}11\text{-}4) - \beta(8\text{-}11\text{-}5)$ $\delta(Cl-O_2)$ ip

$S_{13} = \psi(2\text{-}1\text{-}4) + \psi(2\text{-}1\text{-}5) + \psi(3\text{-}1\text{-}7) + \psi(3\text{-}1\text{-}6) - \psi(3\text{-}1\text{-}4) - \psi(3\text{-}1\text{-}5) - \psi(2\text{-}1\text{-}6) - \psi(2\text{-}1\text{-}7)$ $\rho(CrO_2)$

$S_{14} = r(1\text{-}4) + r(1\text{-}5) - r(1\text{-}6) - r(1\text{-}7)$ $v(Cr-O)$

$S_{15} = \psi(2\text{-}1\text{-}4) + \psi(2\text{-}1\text{-}7) + \psi(3\text{-}1\text{-}4) + \psi(3\text{-}1\text{-}6) - \psi(2\text{-}1\text{-}5) - \psi(3\text{-}1\text{-}5) - \psi(3\text{-}1\text{-}7)$ $\tau(CrO_2)$ op

$S_{16} = \rho(7\text{-}1\text{-}4) + \rho(6\text{-}1\text{-}5)$ $\delta s(O\text{-}Cr\text{-}O)$

$S_{17} = \tau(1\text{-}4\text{-}11\text{-}8) + \tau(1\text{-}4\text{-}11\text{-}13) + \tau(1\text{-}5\text{-}11\text{-}8) + \tau(1\text{-}5\text{-}11\text{-}13) + \tau(1\text{-}6\text{-}10\text{-}9) + \tau(1\text{-}6\text{-}10\text{-}12)$

    $+ \tau(1\text{-}7\text{-}10\text{-}9) + \tau(1\text{-}7\text{-}10\text{-}12)$ $\tau(Cr\text{-}O\text{-}Cl\text{-}O)$ ip

B symmetry

$S_{18} = s(9\text{-}10) - s(11\text{-}8) - s(10\text{-}12) + s(11\text{-}13)$ $va(Cl=O_2)$ op

$S_{19} = s(9\text{-}10) - s(11\text{-}8) + s(10\text{-}12) - s(11\text{-}13)$ $vs(Cl=O_2)$ op

(continued)

Table 3.4 (continued)

$S_{20} = q\,(1\text{-}2) + q\,(1\text{-}3)$	$va\,(Cr{=}O)$
$S_{21} = s\,(10\text{-}6) + s\,(11\text{-}5) - s\,(10\text{-}7) - s\,(11\text{-}4)$	$va\,(Cl\text{-}O_2)$ op
$S_{22} = s\,(10\text{-}6) - s\,(11\text{-}5) + s\,(10\text{-}7) - s\,(11\text{-}4)$	$vs\,(Cl\text{-}O_2)$ op
$S_{23} = 4\alpha\,(6\text{-}10\text{-}7) - 4\alpha(5\text{-}11\text{-}4) - \beta\,(7\text{-}10\text{-}12) - \beta\,(9\text{-}10\text{-}7) - \beta\,(6\text{-}10\text{-}12) - \beta\,(9\text{-}10\text{-}6) + \beta\,(13\text{-}11\text{-}5)$ $+ \beta\,(4\text{-}11\text{-}13) + \beta\,(8\text{-}11\text{-}4) + \beta\,(8\text{-}11\text{-}5)$	$\delta\,(Cl\text{-}O_2)$ op
$S_{24} = \beta\,(7\text{-}10\text{-}12) + \beta\,(9\text{-}10\text{-}7) + \beta\,(13\text{-}11\text{-}5) + \beta\,(8\text{-}11\text{-}5) - \beta\,(4\text{-}11\text{-}13) - \beta\,(8\text{-}11\text{-}4) - \beta\,(6\text{-}10\text{-}12) - \beta\,(9\text{-}10\text{-}6)$	$\rho\,(Cl{=}O_2)$ op
$S_{25} = \beta\,(7\text{-}10\text{-}12) + \beta\,(6\text{-}10\text{-}12) + \beta\,(8\text{-}11\text{-}4) + \beta\,(8\text{-}11\text{-}5) - \beta\,(9\text{-}10\text{-}6) - \beta\,(13\text{-}11\text{-}5) - \beta\,(4\text{-}11\text{-}13) - \beta\,(9\text{-}10\text{-}7)$	wag $(Cl{=}O_2)$ ip
$S_{26} = \beta\,(7\text{-}10\text{-}12) + \beta\,(9\text{-}10\text{-}7) + \beta\,(13\text{-}11\text{-}5) + \beta\,(8\text{-}11\text{-}5) - \beta\,(6\text{-}10\text{-}12) - \beta\,(8\text{-}11\text{-}4) - \beta\,(9\text{-}10\text{-}6) - \beta\,(4\text{-}11\text{-}13)$	$\rho\,(Cl{=}O_2)$ ip
$S_{27} = \tau\,(1\text{-}4\text{-}11\text{-}8) + \tau\,(1\text{-}4\text{-}11\text{-}13) + \tau\,(1\text{-}5\text{-}11\text{-}8) + \tau\,(1\text{-}5\text{-}11\text{-}13) - \tau\,(1\text{-}6\text{-}10\text{-}9)- \tau\,(1\text{-}6\text{-}10\text{-}12) - \tau\,(1\text{-}7\text{-}10\text{-}9)$ $- \tau\,(1\text{-}7\text{-}10\text{-}12)$	$\tau\,(Cr\text{-}O\text{-}Cl\text{-}O)$ op
$S_{28} = \beta\,(7\text{-}10\text{-}12) + \beta\,(9\text{-}10\text{-}6) + \beta\,(4\text{-}11\text{-}13) + \beta\,(8\text{-}11\text{-}5) - \beta\,(13\text{-}11\text{-}5) - \beta\,(6\text{-}10\text{-}12) - \beta\,(9\text{-}10\text{-}7) - \beta\,(8\text{-}11\text{-}4)$	$\tau\,(Cl{=}O_2)$ op
$S_{29} = \psi\,(2\text{-}1\text{-}4) + \psi\,(2\text{-}1\text{-}5) + \psi\,(3\text{-}1\text{-}4) - \psi\,(3\text{-}1\text{-}5) - \psi\,(3\text{-}1\text{-}7) - \psi\,(3\text{-}1\text{-}6) - \psi\,(2\text{-}1\text{-}6) - \psi\,(2\text{-}1\text{-}7)$	wag (CrO_2)
$S_{30} = \psi\,(2\text{-}1\text{-}4) + \psi\,(2\text{-}1\text{-}7) + \psi\,(3\text{-}1\text{-}5) + \psi\,(3\text{-}1\text{-}7) - \psi\,(2\text{-}1\text{-}5) - \psi\,(2\text{-}1\text{-}6) - \psi\,(3\text{-}1\text{-}6) - \psi\,(3\text{-}1\text{-}4)$	$\tau\,(CrO_2)$ ip
$S_{31} = \rho\,(7\text{-}1\text{-}4) - \rho\,(6\text{-}1\text{-}5)$	$\delta a\,(O\text{-}Cr\text{-}O)$
$S_{32} = r\,(1\text{-}4) + r\,(1\text{-}7) - r\,(1\text{-}5) - r\,(1\text{-}6)$	$\nu\,(Cr\text{-}O)$
$S_{33} = r\,(1\text{-}4) + r\,(1\text{-}6) - r\,(1\text{-}5) - r\,(1\text{-}7)$	$\nu\,(Cr\text{-}O)$

q = Cr=O bond distnce, r = Cr-O bond distance, s = Cl-O bond distance, θ = O=Cr=O bond distance, ρ = O-Cr-O bond angle
ϕ = Cr-O-Cl bond angle, ψ = O=Cl=O bond angle, α = O=Cl=O or O-Cl-O bond angles, β = O=Cl-O bond angle
Abbreviations ν stretching, δ deformation, ρ in the plane bending or rocking, γ out-of-plane bending or wagging, τw twisting
a antisymmetric, s symmetric, ip in phase, op out of phase

Table 3.5 Definition of natural internal coordinates for the $C_2(2)$ structure of chromyl perchlorate with the monodentate coordination adopted by perchlorate groups

A symmetry

$S_1 = 2s\,(9\text{-}10) + 2\,s\,(11\text{-}8) - s\,(10\text{-}7) - s\,(10\text{-}12) - s\,(11\text{-}5) - s\,(11\text{-}13)$ — $\nu a\,(ClO_3)$ ip

$S_2 = 2s\,(9\text{-}10) - 2\,s\,(11\text{-}8) - s\,(10\text{-}7) - s\,(10\text{-}12) + s\,(11\text{-}5) + s\,(11\text{-}13)$ — $\nu a\,(ClO_3)$ op

$S_3 = q\,(1\text{-}2) - q\,(1\text{-}3)$ — $\nu a\,(Cr{=}O)$

$S_4 = s\,(9\text{-}10) + s\,(10\text{-}12) + s\,(10\text{-}7) + s\,(11\text{-}5) + s\,(11\text{-}13) + s\,(11\text{-}8)$ — $\nu s\,(ClO_3)$ ip

$S_5 = r\,(1\text{-}4) + r\,(1\text{-}6)$ — $\nu s\,(Cr{-}O)$

$S_6 = \alpha\,(12\text{-}10\text{-}7) + \alpha\,(9\text{-}10\text{-}12) + \alpha\,(9\text{-}10\text{-}7) + \alpha\,(13\text{-}11\text{-}5) + \alpha\,(8\text{-}11\text{-}13) - \beta\,(6\text{-}10\text{-}7)$
$\quad - \beta\,(6\text{-}10\text{-}12) - \beta\,(9\text{-}10\text{-}6) - \beta\,(4\text{-}11\text{-}5) - \beta\,(8\text{-}11\text{-}4) - \beta\,(4\text{-}11\text{-}13)$ — $\delta s\,(ClO_3)$ ip

$S_7 = \alpha\,(9\text{-}10\text{-}12) + \alpha\,(8\text{-}11\text{-}13) - \alpha\,(9\text{-}10\text{-}7) - \alpha\,(8\text{-}11\text{-}5)$ — $\delta a\,(ClO_3)$ ip

$S_8 = 2\alpha\,(12\text{-}10\text{-}7) - \alpha\,(9\text{-}10\text{-}12) - \alpha\,(9\text{-}10\text{-}7) + 2\alpha\,(13\text{-}11\text{-}5) - \alpha\,(8\text{-}11\text{-}5) - \alpha\,(8\text{-}11\text{-}13)$ — $\delta a\,(ClO_3)$ ip

$S_9 = r\,(11\text{-}4) - r\,(10\text{-}6)$ — $\nu a\,(Cl{-}O)$

$S_{10} = \theta\,(2\text{-}1\text{-}3)$ — $\delta\,(O{=}Cr{=}O)$

$S_{11} = 2\beta\,(9\text{-}10\text{-}6) - 2\beta\,(8\text{-}11\text{-}4) - \beta\,(6\text{-}10\text{-}7) - \beta\,(13\text{-}11\text{-}4) - \beta\,(4\text{-}11\text{-}5) - \beta\,(6\text{-}10\text{-}12)$ — $\rho\,(ClO_3)$ ip

$S_{12} = 2\beta\,(9\text{-}10\text{-}6) + 2\beta\,(8\text{-}11\text{-}4) - \beta\,(6\text{-}10\text{-}7) + \beta\,(13\text{-}11\text{-}4) + \beta\,(4\text{-}11\text{-}5) - \beta\,(6\text{-}10\text{-}12)$ — $\rho\,(ClO_3)$ op

$S_{13} = \psi\,(3\text{-}1\text{-}4) + \psi\,(2\text{-}1\text{-}6) - \psi\,(3\text{-}1\text{-}6) - \psi\,(2\text{-}1\text{-}4)$ — $\tau wis\,(CrO_2)$

$S_{14} = \rho\,(6\text{-}1\text{-}4)$ — $\delta\,(O{-}Cr{-}O)$

$S_{15} = \phi\,(1\text{-}6\text{-}10) - \phi\,(1\text{-}4\text{-}11)$ — $\delta a\,(Cr{-}O{-}Cl)$

$S_{16} = \tau\,(3\text{-}1\text{-}4\text{-}11) + \tau\,(2\text{-}1\text{-}4\text{-}11) - \tau\,(2\text{-}1\text{-}6\text{-}10) - \tau\,(3\text{-}1\text{-}6\text{-}10)$ — $\tau\,(OCrOCl)$ op

$S_{17} = \tau\,(1\text{-}4\text{-}11\text{-}5) + \tau\,(1\text{-}4\text{-}11\text{-}8) + \tau\,(1\text{-}4\text{-}11\text{-}13) - \tau\,(1\text{-}6\text{-}10\text{-}7) - \tau\,(1\text{-}6\text{-}10\text{-}9) - \tau\,(1\text{-}6\text{-}10\text{-}12)$ — $\tau\,(ClO_3)$ op

B symmetry

$S_{18} = 2s\,(9\text{-}10) - 2\,s\,(11\text{-}8) - s\,(10\text{-}7) - s\,(10\text{-}12) + s\,(11\text{-}5) + s\,(11\text{-}13)$ — $\nu a\,(ClO_3)$ op

$S_{19} = s\,(11\text{-}13) + s\,(10\text{-}12) - s\,(11\text{-}5) - s\,(10\text{-}7)$ — $\nu a\,(ClO_3)$ ip

$S_{20} = q\,(1\text{-}2) + q\,(1\text{-}3)$ — $\nu s\,(Cr{=}O)$

$S_{21} = s\,(9\text{-}10) + s\,(10\text{-}12) + s\,(10\text{-}7) - s\,(11\text{-}5) - s\,(11\text{-}13) - s\,(11\text{-}8)$ — $\nu s\,(ClO_3)$ op

(continued)

Table 3.5 (continued)

$S_{22} = r\,(1\text{-}4) - r\,(1\text{-}6)$	$\nu a\,(\text{Cr-O})$
$S_{23} = \alpha\,(12\text{-}10\text{-}7) + \alpha\,(9\text{-}10\text{-}12) + \alpha\,(9\text{-}10\text{-}7) - \alpha\,(13\text{-}11\text{-}5) - \alpha\,(8\text{-}11\text{-}5) - \alpha\,(8\text{-}11\text{-}13) - \beta\,(6\text{-}10\text{-}7)$ $- \beta(6\text{-}10\text{-}12) - \beta\,(9\text{-}10\text{-}6) + \beta\,(4\text{-}11\text{-}5) + \beta\,(8\text{-}11\text{-}4) + \beta\,(4\text{-}11\text{-}13)$	$\delta s\,(\text{ClO}_3)\ op$
$S_{24} = 2\alpha\,(9\text{-}10\text{-}12) - 2\alpha\,(13\text{-}11\text{-}5) + \alpha\,(8\text{-}11\text{-}13) - \alpha\,(9\text{-}10\text{-}7) + \alpha\,(8\text{-}11\text{-}5) - \alpha\,(9\text{-}10\text{-}12)$	$\delta a\,(\text{ClO}_3)\ op$
$S_{25} = \alpha\,(9\text{-}10\text{-}12) + \alpha\,(8\text{-}11\text{-}5) - \alpha\,(9\text{-}10\text{-}7) - \alpha\,(8\text{-}11\text{-}13)$	$\delta a\,(\text{ClO}_3)\ op$
$S_{26} = r\,(11\text{-}4) + r\,(10\text{-}6)$	$\nu s\,(\text{Cl-O})$
$S_{27} = \psi\,(2\text{-}1\text{-}4) + \psi\,(3\text{-}1\text{-}4) - \psi\,(2\text{-}1\text{-}6) - \psi\,(3\text{-}1\text{-}6)$	$wag\,(\text{CrO}_2)$
$S_{28} = \beta\,(6\text{-}10\text{-}7) + \beta\,(13\text{-}11\text{-}4) - \beta\,(4\text{-}11\text{-}5) - \beta\,(6\text{-}10\text{-}12)$	$\rho\,(\text{ClO}_3)\ ip$
$S_{29} = \beta\,(6\text{-}10\text{-}7) - \beta\,(13\text{-}11\text{-}4) + \beta\,(4\text{-}11\text{-}5) - \beta\,(6\text{-}10\text{-}12)$	$\rho\,(\text{ClO}_3)\ op$
$S_{30} = \psi\,(3\text{-}1\text{-}6) + \psi\,(3\text{-}1\text{-}4) - \psi\,(2\text{-}1\text{-}4) - \psi\,(2\text{-}1\text{-}6)$	$\rho\,(\text{CrO}_2)$
$S_{31} = \phi\,(1\text{-}6\text{-}10) + \phi\,(1\text{-}4\text{-}11)$	$\delta s\,(\text{Cr-O-Cl})$
$S_{32} = \tau\,(1\text{-}3\text{-}4\text{-}11) + \tau\,(2\text{-}1\text{-}4\text{-}11) + \tau\,(2\text{-}1\text{-}6\text{-}10) + \tau\,(3\text{-}1\text{-}6\text{-}10)$	$\tau\,(\text{OCrOCl})\ ip$
$S_{33} = \tau\,(1\text{-}4\text{-}11\text{-}5) + \tau\,(1\text{-}4\text{-}11\text{-}8) + \tau\,(1\text{-}4\text{-}11\text{-}13) + \tau\,(1\text{-}6\text{-}10\text{-}7) + \tau\,(1\text{-}6\text{-}10\text{-}9) + \tau\,(1\text{-}6\text{-}10\text{-}12)$	$\tau\,(\text{ClO}_3)\ ip$

$q = \text{Cr=O}$ bond distance, $r = \text{Cr-O}$ bond distance, $s = \text{Cl-O}$ bond distance, $\theta = \text{O=Cr=O}$ bond distance, $\rho = \text{O-Cr-O}$ bond angle, $\rho = \text{O-Cr-O}$ bond angle
$\phi = \text{Cr-O-Cl}$ bond angle, $\psi = \text{O=Cr-O}$ bond angles, $\alpha = \text{O=Cl-O}$ bond angle, $\beta = \text{O=Cl-O}$ bond angle
Abbreviations ν stretching, δ deformation, ρ in the plane bending or rocking, γ out-of-plane bending or wagging, τw twisting
a antisymmetric, s symmetric, ip in phase, op out of phase

perchlorate group of the $C_2(1)$ structure and monodentate for the perchlorate group of the $C_2(2)$ structure).

These results are supported by the highly strong band in the 1000–900 cm^{-1} region and also by the intense bands observed at lower frequencies (490 cm^{-1}). The intensities of these bands (not predicted by calculations) are probably due to overlapping between the Cr=O, Cl=O, and Cl–O stretching and deformation modes of both coordination modes of the two structures. A comparison between the average calculated spectra (from B3P86/6-31G* and B3LYP/6-31G* levels for $C_2(1)$ and $C_2(2)$ structures, respectively) using average frequencies and intensities with the corresponding experimental demonstrate a good correlation, as can be seen in Fig. 6 of Ref. [2].

3.4 Force Field

The harmonic force fields in Cartesian coordinates were turned into the local symmetry or "natural" coordinates proposed by Fogarasi et al. [19], which are given from Tables 3.3 to 3.5. The force constants for chromyl perchlorate were estimated using the scaling procedure of Pulay et al. [20]. The scaling factors affecting the main force constants were subsequently calculated by an interactive procedure using the MOLVIB program [21, 22]. The resulting numbers for the three cases considered are collected in Tables 12 and 13 of Ref. [2].

Frequencies, infrared intensities, Raman activities, and PED obtained from chromyl perchlorate appear in Tables 8, 9, and 11 of Ref. [2], for the two coordination modes considered for the perchlorate groups. The force constants appearing in Tables 14 and 15 of Ref. [2] expressed in terms of simple valence internal coordinates, which were calculated from the corresponding scaled force fields.

3.5 Conclusions

In this chapter a structural and vibrational study on the chromyl perchlorate is presented. Two structures with C_2 symmetries, named $C_2(1)$ and $C_2(2)$ were obtained using ab initio and DFT calculations. Here, an approximate normal coordinate analysis, considering the mode of coordination adopted by perchlorate groups as monodentate and bidentate for the $C_2(1)$ structure and monodentate for the $C_2(2)$ structure of chromyl perchlorate, was presented. Both structures may be present in the compound in the solid phase because the vibrational and force constants analyses suggest that both coordinations represent the perchlorate group in chromyl perchlorate due to two structures.

The molecular force field and vibrational frequencies were obtained using the B3P86/6-31G* method for the $C_2(1)$ structure and the B3LYP/6-31G* level for the $C_2(2)$ structure.

The assignments previously made [18] were corrected and completed in accordance with the theoretical results. The assignments of the 33 normal vibration modes corresponding to chromyl perchlorate are reported.

The hexacoordination of the Cr atom in chromyl perchlorate for the $C_2(1)$ structure of chromyl perchlorate and the coordination monodentate for the $C_2(2)$ structure were confirmed using the Wiberg's indexes calculated by means of the natural bond orbital (NBO) study and AIM analyses.

Acknowledgments This work was subsidized with grants from CIUNT (Consejo de Investigaciones, Universidad Nacional de Tucumán). The author thanks Prof. Tom Sundius for his permission to use MOLVIB.

References

1. S.A. Brandán, J. Mol. Struc. (THEOCHEM) **908**,19 (2009)
2. S. A. Brandán, *Structural and vibrational properties of chromyl perchlorate*, ed. by L. E. Mattews, Chapter 3, (Nova Science Publisher, Inc, Hauppauge, 2010)
3. S.A. Brandán, M.L. Roldán, C. Socolsky, A. Ben Altabef, Spectrochim. Acta, Part A **69**, 1027 (2008)
4. A. Ben Altabef, S.A. Brandán, J. Mol. Struc. **981**, 146 (2010)
5. C.J. Marsden, K. Hedberg, M.M. Ludwig, G.L. Gard, Molecular structure of $CrO_2(NO_3)_2$ in gas phase: a novel form of coordination. Inorg. Chem. **30**, 4761–4766 (1991)
6. R.J. Gillespie (ed.), *Molecular Geometry* (Van Nostrand-Reinhold, London, 1972)
7. R.J. Gillespie, I. Bytheway, T.H. Tang, R.F.W. Bader, Geometry of the fluorides, oxofluorides, and methanides of vanadium (V), chromium (VI), and molybdenum (VI): understanding the geometry of non-VsEpR molecules in terms of core distortion. Inorg. Chem. **35**, 3954–3963 (1996)
8. S. A. Brandán, M. L. Roldán, C. Socolsky, A. Ben Altabef, DFT calculation of the chromyl nitrate, $CrO_2(NO_3)_2$:the molecular force field. Spectrochim. Acta A, **69**, 1027–1043 (2008)
9. A.E. Reed, L.A. Curtis, F. Weinhold, Chem. Rev. **88**(6), 899 (1988)
10. R.F.W. Bader, A bond path: a universal indicator of bonded interactions. J. Phys. Chem. A **102**, 7314–7323 (1998)
11. F. Biegler-Köning, J. Schönbohm, D. Bayles, AIM2000; A program to analyze and visualize atoms in molecules. J. Comput. Chem. **22**, 545–559 (2001)
12. S. Wojtulewski, S.J. Grabowski, DFT and AIM studies on two ring resonance assisted hydrogen bonds. J. Mol. Struct. **621**, 285–291 (2003)
13. S.J. Grabowski, Properties of a ring critical point as measure of intramolecular H-bond strength. Monat. für Chem. **133**, 1373–1380 (2002)
14. C. C. Addison, N. Logan, S. C. Wallwork, C. D. Garner, *Q.* Structural aspects of coordinated nitrate groups. Rev. Chem. Soc, **25**, 289–322 (1971)
15. C. C. Addison, N. Logan, Anhydrous metal nitrates. Adv. Inorg. Chem. Radiochim., **6**, 71–142 (1964)
16. C. C. Addison, D. Sutton, Complexes containing the nitrate ion. Prog. Inorg. Chem., **8**, 195–276 (1967)
17. J. Laane, J.R. Ohlsen, Characterization of nitrogen oxides by vibrational spectroscopy. Prog. Inorg. Chem. **27**, 465–511 (1980)
18. M. Chaabouni, T. Chausse, J.L. Pascal, J. Potier, Synthesis and study of two perchlrorates of crhomium (III). J. Chem. Res. **5**, 72–73 (1980)

19. G. Fogarasi, P. Pulay, in *Vibrational Spectra and Structure,* vol. 14, ed. by J. E. Durig, (Elsevier, Amsterdam, 1985), p. 125
20. P. Pulay, G. Fogarasi, F. Pang, J.E. Boggs, Systematic ab initio gradient calculation of molecular geometries, force constants and dipole moment derivatives. J. Am. Chem. Soc. **101**(10), 2550–2560 (1979)
21. T. Sundius, A computer program for normal coordinate treatment of molecular vibrations. J. Mol. Struct. **218**, 321–326 (1990)
22. T. Sundius, MOLVIB: a program for harmonic force field calculation, QCPE program no. 604, 1991

References

Chapter 4
Theoretical Structural and Vibrational DFT Calculations of Chromyl Thiocyanate

Abstract In this chapter, the structural and vibrational properties of chromyl thiocyanate were studied using density functional theory (DFT) methods. The initial geometries were fully optimized at different theory levels and the harmonic wavenumbers were evaluated at the same levels. Also, the characteristics and nature of the Cr–O and Cr ← O bonds for the stable structure were studied by means of the natural bond orbital (NBO) study while the topological properties of electronic charge density were analyzed using *Bader*'s atoms in the molecules theory (AIM). Besides, a complete assignment of all observed bands in the infrared spectrum for the compound was performed combining DFT calculations with Pulay's scaled quantum mechanics force field (SQMFF) methodology.

Keywords Chromyl thiocyanate · Vibrational spectra · Molecular structure · Force field · DFT calculations

4.1 Introduction

The chromyl compounds are interesting from the structural and spectroscopic points of view because their vibrational properties and reactivities are strongly dependent on the coordination modes that adopt the different ligands linked to the chromyl groups and also on the stereochemistry of these compounds [1–5]. Particularly, the physics and chemical properties of inorganic and organic thiocyanates were reported by numerous authors as well as structural studies of different thiocyanates [6–13]. So far, there is no theoretical study concerning either geometry or vibrational spectra for chromyl thiocyanate and for this here, a comparative work was performed intended to evaluate the best level of theory and basis set to reproduce the experimental data existing for chromyl thiocyanate, in order to predict the molecular geometry and the

S. A. Brandán, *A Structural and Vibrational Investigation into Chromylazide, Acetate, Perchlorate, and Thiocyanate Compounds*, SpringerBriefs in Molecular Science, DOI: 10.1007/978-94-007-5754-7_4, © The Author(s) 2013

Fig. 4.1 The C_2 molecular
structure of chromyl
thiocyanate

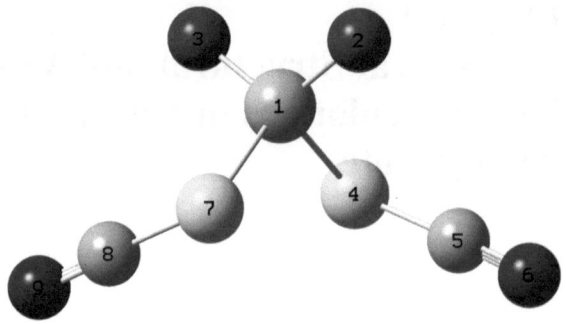

vibrational spectra for which no experimental data exist. For this purpose, the optimized geometries and wavenumbers for the normal modes of vibration of the sodium sulfocyanide [7], sulfocyanide anion and chromyl sulfocyanide compounds were calculated. The harmonic force constants given by these calculations were subsequently scaled to reproduce as well as possible the proposed experimental wavenumbers obtained from those similar molecules. Then, the obtained results were compared and analyzed. In addition, the nature of the different types of bonds and their corresponding topological properties of electronic charge density were systematic and quantitatively investigated, for those compounds, by using the natural bond orbital (NBO) [14–16] analysis and the atoms in molecules theory (AIM) [17].

4.2 Structural Analysis

The initial geometry of different thiocyanate compounds were modeled with the *Gaussian View* program [18] while the geometry calculations were performed using the Gaussian 03 programs [19]. Calculations were made with the hybrid B3LYP method [20, 21] as implemented in the Gaussian programs. The Lanl2dz, STO-3G*, 6-31G*, 6-31 + G*, 6-311G*, and 6-311 + G* basis sets were used. For chromyl thiocyanate a stable structure of C_2 symmetry was found, as observed in Fig. 4.1.

The natural coordinates for the compound are shown in Table 4.1 and were defined as proposed by Fogarasi et al. [22]. The numbering of the atoms for chromyl thiocyanate is described in Fig. 4.1.

The normal mode analysis for the structure of chromyl thiocyanate was carried out with the B3LYP/6-31G* method. The harmonic force field in cartesian coordinates for the compound which resulted from the calculations were transformed into "natural" internal coordinates [23] by the MOLVIB program [24, 25]. The potential energy distribution (PED) components larger than or equal to 10 % are subsequently calculated with the resulting scaled quantum mechanics (SQM) force field. The NBO analysis was then performed using the B3LYP/6-31G* level with the NBO 3.1 program [26] included in Gaussian 03 package [19].

Table 4.1 Definition of natural internal coordinates for chromyl thiocyanate

A symmetry	
$S_1 = s$ (5–6) $+ s$ (8–9)	vip (C–N)
$S_2 = s$ (4–5) $+ s$ (7–8)	vip (S–C)
$S_3 = q$ (1–2) $+ q$ (1–3)	v_sCr=O
$S_4 = \phi$ (4–5–6) $- \phi$ (7–8–9)	δS–C–N op
$S_5 = \tau$ (1–4–5–6) $+ \tau$ (1–7–8–9)	τCr–S–C–N ip
$S_6 = r$ (1–4) $- r$ (1–7)	v_aCr–S
$S_7 = r$ (1–4) $+ r$ (1–7)	v_sCr–S
$S_8 = \psi$ (3–1–6) $+ \psi$ (3–1–4) $- \psi$ (2–1–4) $- \psi$ (2–1–6)	ρCrO$_2$
$S_9 = \alpha$ (9–10–12)	δS–Cr–S
$S_{10} = \tau$ (3–1–4–11) $+ \tau$ (2–1–4–11) $- \tau$ (2–1–6–10) $- \tau$ (3–1–6–10)	τO–Cr–S–C ip
$S_{11} = \tau$ (3–1–4–11) $+ \tau$ (2–1–4–11) $- \tau$ (2–1–6–10) $- \tau$ (3–1–6–10)	τO–Cr–S–C op
B symmetry	
$S_{12} = s$ (5–6) $- s$ (8–9)	vop (C–N)
$S_{13} = s$ (4–5) $- s$ (7–8)	vop (S–C)
$S_{14} = q$ (1–2) $- q$ (1–3)	v_aCr=O
$S_{15} = \phi$ (4–5–6) $+ \phi$ (7–8–9)	δS–C–N ip
$S_{16} = \tau$ (1–4–5–6) $- \tau$ (1–7–8–9)	τCr–S–C–N op
$S_{17} = \theta$ (2–1–3)	δCrO$_2$
$S_{18} = \psi$ (2–1–4) $+ \psi$ (3–1–4) $- \psi$ (2–1–6) $- \psi$ (3–1–6)	wagCrO$_2$
$S_{19} = \psi$ (3–1–4) $+ \psi$ (2–1–6) $- \psi$ (3–1–6) $- \psi$ (2–1–4)	τCrO$_2$
$S_{20} = \beta$ (1–4–5) $+ \alpha$ (1–7–8)	δCr–S–C ip
$S_{21} = \beta$ (1–4–5) $- \alpha$ (1–7–8)	δCr–S–C op

$q =$ Cr=O bond distance, $r =$ Cr–N bond distance, $s =$ N–N bond distance, $\theta =$ O=Cr=O bond angle, $\alpha =$ N–Cr–N bond angle, $\phi =$ N–N–N bond angle, $\psi =$ O=Cr–N bond angles, $\beta =$ Cr–N–N bond angle

Abbreviations v stretching, δ deformation, ρ in the plane bending or rocking, wag out- of-plane bending or wagging, τ torsion, a antisymmetric, s symmetric, ip in phase, op out of phase

The topological properties of the compound charge density were computed at the same theory level with the AIM2000 software [27].

A comparison of the total energies and dipole moment values for chromyl thiocyanate with the corresponding values for sodium and chloro thiocyanate and thiocyanate ion using the Lanl2dz, STO-3G*, 6-31G*, 6-31 + G*, 6-311G*, and 6-311 + G* basis sets with the B3LYP method are shown in Table 4.2. The optimized structure for the chromyl thiocyanate has C_2 symmetry; the corresponding structures for sodium and chloro thiocyanate have C_S symmetries, while for the thiocyanate ion the structure has $C_{\infty V}$ symmetry.

For chromyl and sodium thiocyanate, the highest dipole moment values are obtained using Lanl2dz basis set, while the most stable structure for all compounds with the B3LYP/6-311 + G* method are obtained. For chloro thiocyanate and thiocyanate ion the highest dipole moment values are obtained using the 6-311 + G* basis set. A comparison of the experimental data for chromyl chloride [28], fluoride [29], nitrate [5], and azide with the calculated geometrical parameters for chromyl thiocyanate using a 6-31G* basis set can be seen in Table 4.3.

Table 4.2 Total energy (ET) and dipole moment (μ) for chromyl thiocyanate, sodium thiocyanate, chloro thiocyanate, and thiocyanate ion using different basis sets

B3LYP method

Basis set	$CrO_2(SCN)_2$		$Na(SCN)$		$Cl(SCN)$		SCN^-	
	C_2 symmetry		C_S symmetry		C_S symmetry		$C_{\infty v}$ symmetry	
	ET (Hartree)	μ (D)	ET (Hartree)	μ (D)	ET (Hartree)	μ (D)	ET (Hartree)	μ (D)
LanL2DZ	−442.584570	1.93	−103.178141	9.36	−117.816128	2.17	−102.997319	1.97
STO-3G*	−2153.049378	1.41	−645.827470	4.86	−940.658406	2.55	−488.719344	0.55
6-31G*	−2176.949214	1.34	−653.382208	7.87	−951.192381	2.30	−491.100554	1.52
6-31 + G*	−2176.976746	1.45	−653.388593	8.15	−951.199266	2.55	−491.120774	1.98
6-311G*	−2177.142700	1.59	−653.440503	8.11	−951.269799	2.38	−491.159996	1.41
6-311 + G*	−2177.172884	1.50	−653.444515	8.12	−951.275983	2.58	−491.169664	2.01

Table 4.3 Comparison of calculated geometrical parameters for chromyl azide compared with the corresponding experimental values for similar compounds

Parameter	C_2 symmetry			C_{2V} symmetry	
	$CrO_2(SCN)_2$[a]	$CrO_2(N_3)_2$[b]	$CrO_2(NO_3)_2$[c]	CrO_2Cl_2[d]	CrO_2F_2[e]
$r(Cr–O)$ (Å)	1.550	1.555	1.586	1.577	1.572
$r(Cr–X)$ (Å)	2.228	1.835	1.957	2.122	1.716
RMSD	–	0.07	0.008	0.006	0.13
$\theta(OCrO)$ (°)	113.0	111.8	112.6	108.5	107.8
$\theta(OCrX)$ (°)	104.0	107.9	104.5	108.7	109.3
$\theta(XCrX)$ (°)	113.4	109.3	140.4	113.2	111.9
RMSD	–	2.8	0.45	4.6	5.2

[a] This work B3LYP/6-31G*
[b] B3LYP/6-31G*
[c] Ref [5] B3LYP/6-31G*
[d] Ref [28]
[e] Ref [29]

According to these results, the experimental geometrical parameters that best reproduce the compound with C_2 structure are those corresponding to chromyl chloride and nitrate with C_{2V} structure where the mean difference for bond lengths is 0.006 Å, while for angles it is 0.45°, respectively.

The B3LYP/6-31G* calculations predict for chromyl thiocyanate approximately a same value for the $O_2=Cr_1=O_3$ bond angle in relation to the $O4–Cr1–O8$ angle, in accordance with the valence-shell electron pair (VSEPR) theory [30, 31].

A comparison of the experimental data for chromyl azide and other thiocyanates [32] with the calculated geometrical parameters for chromyl thiocyanate using a 6-31G* basis set as shown in Table 4.4. The experimental values for the C–S and C–N distances are, respectively, 1.59 and 1.25 Å [32].

Table 4.4 Comparison of calculated geometrical parameters for chromyl thiocyanate compared with the corresponding experimental values for similar thiocyanates

Parameter	C_2 symmetry		C_S symmetry		
	$CrO_2(SCN)_2$[a]	$CrO_2(N_3)_2$[b]	HSCN[c]	NaSCN[c]	Cl(SCN)[c]
r(S–C) (Å)	1.697	1.231	1.706	1.670	1.694
r(C–N) (Å)	1.167	1.140	1.165	1.184	1.167
RMSD	–	1.18	0.006	0.02	0.002
θ(SCN) (°)	177.8	174.4	176.6	168.4	175.5
θ(MSC) (°)	99.6	125.5	95.6	56.60	101.7
RMSD	–	18.5	2.9	44.6	2.2

[a] This work B3LYP/6-31G*
[b] B3LYP/6-31G*
[c] Ref [32]

Table 4.5 Natural population atomic (NPA) and Wiberg Index of chromyl thiocyanide B3LYP/6-31G*

Numbers	Atoms	NPA	Bond order
1	Cr	1.261	5.513
2	O	−0.375	2.363
3	O	−0.375	2.363
4	S	−0.009	2.409
5	C	0.002	4.005
6	N	−0.248	3.053
7	S	−0.009	2.409
8	C	0.002	4.005
9	N	−0.248	3.053

The values calculated for chromyl thiocyanate clearly show a good agreement with the experimental values observed for the bond distances and angles of Cl(SCN) [32].

4.3 NBO Analysis

The bond orders, expressed by Wiberg's indexes for chromyl thiocyanate are given in Table 4.5 together with the atomic charges values (NPA). In this case, the chromium atom forms four bonds, two Cr=O bonds and two Cr–S. This analysis clearly shows the monodentate coordinations for the SCN groups of chromyl thiocyanate.

Table 4.6 Analysis of the bond critical points in chromyl thiocyanate compared with similar molecules*

B3LYP/6-31G*Method

Parameter	$CrO_2(SCN)_2$[a]		CrO_2Cl_2[a]	CrO_2F_2[a]	NaSCN[a]	HSCN[a]	ClSCN[a]
	Cr1-S4	Cr1-S7	Cr-Cl	Cr-F	Na-S	H-S	Cl-S
$\rho(r)$	0.09701	0.09701	0.10851	0.17525	0.02207	0.20466	0.13042
$\nabla^2\rho(r)$	0.16229	0.16205	0.29701	1.12065	0.14020	−0.49613	−0.04120
$\lambda 1$	−0.13594	−0.13617	−0.15099	−0.36390	−0.02423	−0.42988	−0.17189
$\lambda 2$	−0.11229	−0.11229	−0.14685	−0.36085	−0.01213	−0.39232	−0.15546
$\lambda 3$	0.41053	0.41051	0.59486	1.81540	0.17657	0.32607	0.28615
$|\lambda 1|/\lambda 3$	0.33113	0.33170	0.25382	0.20045	0.13722	1.31836	0.60069
[b]$M-X$ (Å)	1.835	1.835	2.126	1.698	2.045	1.023	
[c]χ^E	Cr = 1.66	N = 3.04	Cl = 3.16	F = 3.98	Na = 0.93	H = 2.20	
[d]M (uma)	Cr = 51.9960	N = 14.0067	Cl = 35.453	F = 18.9980	Na = 22.9897	H = 1.0079	

* the quantities are in atomic units

[a] This work

[b] (M=Cr, Na; X=N, Cl, F)

[c] X^E : electronegativity of X

[d] M: Cr, N, Cl, F, Na, H

4.4 AIM Analysis

The localization of the critical points in the charge electron density, $\rho(r)$ and the values of the Laplacian, $\nabla^2\rho(r)$ at these points are important to characterize the molecular electronic structure of a compound in terms of the magnitude and interactions nature. The analyses of the bonds' critical points in the compound studied are reported with the B3LYP/6-31G* method in Table 4.6. In this case, a very important observation is the different nature of the Cr–S bonds in chromyl thiocyanate in relation to the Cr–Cl or Cr–F bonds in the chromyl chloride and fluoride compounds, respectively. The topological properties have lower values in the first case, while the $\rho(r)$ value is bigger in chromyl fluoride according to the electronegativity of the F atom and to a lower value in the Cr–F distance (1.698 Å). On the other hand, the low value of $\rho(r)$ in the sodium thiocyanate shows clearly an ionic nature of this compound in reference to the other thiocyanate compounds. However, the chromyl thiocyanate has a low behavior ionic compared with the sodium thiocyanate. Finally, these results confirm that both thiocyanate groups in chromyl thiocyanate act with a monodentate coordination.

4.5 Vibrational Analysis

The structure of chromyl thiocyanate has C_2 symmetry and 21 active vibrational normal modes, classified as: $11\ A + 10\ B$. The calculated frequencies and the assignment for chromyl thiocyanate are given in Table 4.7. The theoretical infrared and Raman spectra compared with those corresponding to chloro thiocyanate and chromyl chloride are given in Figs. 4.2 and 4.3.

Vibrational assignments were made on the basis of the potential energy distributions (PED) in terms of symmetry coordinates and by comparison with molecules that contain similar chromyl and thiocyanate groups [1–4, 33]. Table 4.8 shows the calculated SQM harmonic frequencies for chromyl thiocyanate using B3LYP method with the 6-31G* basis set and the PED contributions. Below, we discuss the assignment of the most important groups for the compounds studied (see Table 4.7).

4.6 Chromyl Groups

In chromyl compounds, the antisymmetric and symmetric Cr=O stretchings appear in the 1050–900 cm^{-1} region [1–5]. In this compound, the frequencies predicted for these vibrational modes show that the stretching modes are split by 17 cm^{-1}, indicating a slight contribution of the central Cr atom in these vibrations. The antisymmetric and symmetric Cr=O stretching modes were calculated in the

Table 4.7 Calculated frequencies (cm^{-1}), potential energy distribution, and assignment for chromyl thiocyanate

Modes	Calculated[b]	IR[c]	Ra[d]	SQM[e]	Assignment[a]
A					
1	2265	0.1	568.3	2165	v_s(C–N)
2	1125	94.7	37.2	989	v_sCr=O
3	696	0.5	19.9	666	v_s(S–C)
4	463	2.9	31.9	465	δCrO$_2$
5	403	0.9	7.9	364	δS–C–Nip
6	388	2.4	2.5	348	τCr–S–C–Nip
7	311	0.8	11.0	304	v_sCr–S
8	209	0.8	7.6	251	ρCrO$_2$
9	137	0.2	6.6	137	δS–Cr–S
10	81	2.2	9.2	79	δCr–S–Cip
11	38	1.9	9.7	39	τO–Cr–S–Cip
B					
12	2266	3.8	177.9	2166	v_a(C–N)
13	1142	105.6	11.1	1004	v_aCr=O
14	696	1.5	6.4	666	v_a(S–C)
15	479	77.1	3.3	507	v_aCr–S
16	391	1.9	0.6	355	τO–Cr–S–Cop
17	383	20.9	0.7	336	δS–C–Nop
18	231	2.2	2.2	273	wagCrO$_2$
19	187	0.2	7.1	224	τCrO$_2$
20	98	3.7	3.0	96	τO–Cr–S–Cop
21	53	4.1	2.9	55	δCr–S–Cop

[a] This work
[b] DFT B3LYP/6-31G*
[c] Units are km mol^{-1}
[d] Raman activities in Å^4 (amu)$^{-1}$
[e] From scaled quantum mechanics force field

infrared spectrum at 1142 and 1125 cm^{-1}, respectively, where the last band is more intense in the Raman spectrum, as expected. As observed in Table 4.8 and in similar molecules [1–5], these modes are calculated uncoupled with other modes. Taking into account the difference between these modes, they are assigned, as in CrO$_2$Cl$_2$, at 1002 and 991 cm^{-1}, thus, the scaled frequencies for both modes decrease until 1004 and 989 cm^{-1}, respectively, with a difference between them of 15 cm^{-1}. Table 4.7 shows for chromyl thiocyanate that the unscaled density functional theory (DFT) frequencies for the antisymmetric Cr=O stretching mode, are higher than the frequencies of the symmetric Cr=O stretching. On the other hand, the antisymmetric and symmetric Cr–S stretchings are split by more than 168 cm^{-1}, indicating a strong contribution of the central Cr atom in these vibrations. In this case, the theoretical calculations predict these modes with greater PED (72 %) value for the antisymmetric mode in reference to the symmetric mode (38 %). In this way, they are assigned at 507 and 304 cm^{-1} with a difference

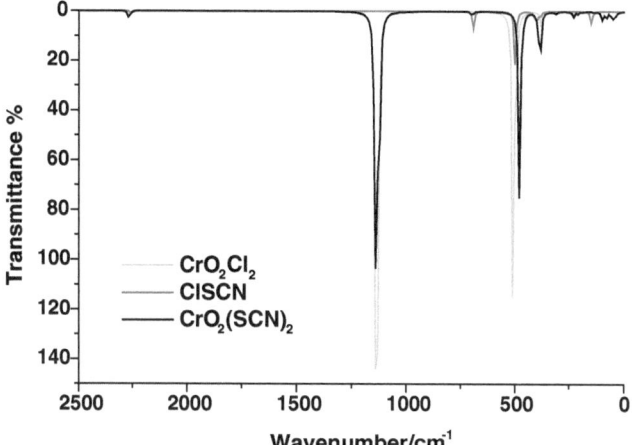

Fig. 4.2 Theoretical infrared spectra of chromyl thiocyanate in the 2500 to 0 cm^{-1} region compared with those corresponding to chloro thiocyanate and chromyl chloride

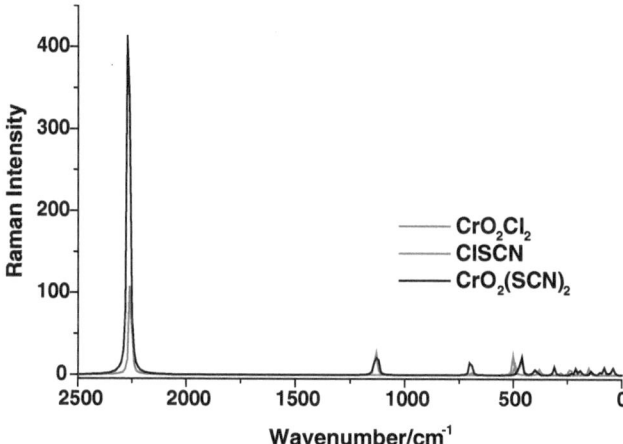

Fig. 4.3 Theoretical Raman spectra of chromyl thiocyanate in the 2500 to 0 cm^{-1} region compared with those corresponding to chloro thiocyanate and chromyl chloride

between both scaled modes of 203 cm^{-1}. In CrO_2F_2 and CrO_2Cl_2, the CrO_2 bending modes are observed at 403 and 356 cm^{-1}, respectively [28, 29]. In this chapter, the B3LYP/6-31G* method calculates the CrO_2 bending at 463 cm^{-1}. For the chromyl compounds, in general, the calculations predict the wagging, rocking, and twisting modes of the CrO_2 group in the low frequencies region and coupled with other modes of the thiocyanate groups. The CrO_2 wagging and rocking modes are assigned to the bands at 273 and 251 cm^{-1}, respectively. As indicated the PED values, the CrO_2 twisting mode is calculated strongly coupled with vibrations of

Table 4.8 Calculated SQM frequencies for chromyl thiocyanate using the B3LYP/6-31G* method together with the PED contributions

Symmetry	IR[c]	SQM[b]	PED $\geq$ 10 %[a]
B	2162	2165	S_1 (93)
A	2162	2165	S_2 (93)
B	1002	1004	S_6 (98)
A	991	989	S_5 (97)
B	678	668	S_3 (92)
A	678	668	S_4 (92)
B	503	508	S_8 (72), S_{17} (22)
A	470	465	S_7 (53), S_{11} (34)
A	356	365	S_{11} (41), S_{20} (17), S_9 (16)
B	353	355	S_{21} (59), S_{14} (14), S_{10} (12)
A	353	347	S_{20} (69)
B	334	337	S_{21} (33), S_{10} (33), S_{17} (15), S_{14} (10)
A	311	303	S_7 (38), S_9 (28), S_{11} (22)
B	268	272	S_{17} (59), S_8 (28), S_{10} (12)
A	257	250	S_{18} (82), S_9 (12)
B	224	225	S_{19} (76), S_{10} (10)
A	139[d]	137	S_{12} (73), S_{16} (11)
B	98[d]	96	S_{15} (36), S_{14} (32), S_{10} (24)
A	81[d]	78	S_{13} (59), S_9 (27)
B	53[d]	55	S_{15} (53), S_{14} (25)
A	38[d]	40	S_{16} (77)

[a] This work
[b] From scaled quantum mechanics force field
[c] From Ref. [28]
[d] Theoretical values using DFT B3LYP/6-31G*

the same group and the thiocyanate groups (Table 4.8). This mode could be assigned at 224 cm^{-1} because it appears with a higher PED value. The S–Cr–S bending mode is calculated at 137 cm^{-1} and assigned at this wavenumber, as observed in Table 4.8.

4.7 Thiocyanate Groups

According to Nakanishi and Solomon [33], the antisymmetric and symmetric C$\equiv$N stretching modes of the thiocyanate groups are observed at 2140 cm^{-1}, for this, and in accordance with our calculations those modes are assigned at 2166 and 2165 cm^{-1}. On the other hand, the antisymmetric and symmetric S–C stretching modes are observed by those authors between 800 and 600 cm^{-1}. Thus, in chromyl thiocyanate both modes are assigned at 666 cm^{-1}, as predicted by calculations and, as observed in Tables 4.7 and 4.8.

Table 4.9 Scale factors for the force field of chromyl thiocyanate

Coordinates	$CrO_2(SCN)_2$
$\nu C \equiv N$. $\nu S-C$	0.91585
$\nu Cr=O$	0.76559
$\nu Cr-S$	1.12620
$\delta S-C-N$	0.56857
δCrO_2	0.99457
$\delta S-Cr-S$	1.02416
$\tau O-Cr-S-C$	1.06362
$wag CrO_2$. ρCrO_2. τCrO_2	1.58834
$\tau Cr-S-C-N$	0.78310

ν stretching, δ deformation, ρ rocking, wag wagging, τ torsion, and theoretical frequencies

Table 4.10 Force constants of chromyl thiocyanate

Coordinates	$CrO_2(SCN)_2$
$f(C \equiv N)$	17.13
$f(S-C)$	3.94
$f(Cr=O)$	7.00
$f(Cr-S)$	2.26
$f(S-C-N)$	0.18
$f(O-Cr-O)$	1.36
$f(S-Cr-S)$	0.58
$f(O-Cr-S-C)$	0.16
$f(wag CrO_2)$	0.93
$f(\rho CrO_2)$	0.80
$f(\tau CrO_2)$	0.90

Units mdyn $Å^{-1}$ for stretching and mdyn $Å$ rad^{-2} for bond deformation

Note that the theoretical infrared and Raman spectra for chromyl chloride and chloro thiocyanate compounds are similar to that corresponding to the chromyl thiocyanate, as shown in Figs. 4.2 and 4.3. Thus, the experimental bands of those compounds are proposed as experimental bands for chromyl thiocyanate in order to realize a complete assignment of the compound by means of the scaled quantum mechanics force field (SQMFF) method. With the exception of the bands in the region between 139 and 38 cm^{-1}, where the calculated frequencies were considered as experimental because they were not observed in the vibrational spectrum of those similar molecules. In this case, the scale factors for the force field of chromyl thiocyanate can be seen in Table 4.9.

4.8 Force Field

The force constants for chromyl thiocyanate were estimated using the scaling procedure of *Pulay* et al. [23], as mentioned before. The harmonic force fields in cartesian coordinates were transformed to the local symmetry or "natural" coordinates proposed by Fogarasi et al. [22], as defined in Table 4.1 (See Fig. 4.1). The scaling factors affecting the main force constants were subsequently calculated by an iterative procedure [24, 25] to have the best possible fit between those proposed as experimental and theoretical frequencies. The resulting numbers were collected in Table 4.9. The main force constants for chromyl thiocyanate appearing in Table 4.10 and the results show values similar to that observed for CrO_2F_2 and CrO_2Cl_2. In those compounds, the scaled GVFF force constants (B3LYP/Lanl2DZ method) for the Cr=O stretchings are 7.443 and 7.122 mdyn $Å^{-1}$, respectively, while the corresponding force constants for the O=Cr=O deformations are 1.110 and 0.938 mdyn $Å$ rad^{-2} [19].

4.9 Conclusions

In this chapter, an approximate normal coordinate analysis for chromyl thiocyanate was proposed together at the vibrational spectra and the assignment complete of the 21 normal modes of vibration.

An SQM force field was obtained for chromyl thiocyanate after adjusting the theoretically obtained force constants in order to minimize the difference between proposed and calculated frequencies.

Here, a DFT molecular force field for the chromyl azide computed using B3LYP/6-31G* is well represented.

The NBO and AIM studies confirm the monodentate coordination of the thiocyanate groups in chromyl thiocyanate.

Acknowledgments This work was subsidized with grants from CIUNT (Consejo de Investigaciones, Universidad Nacional de Tucumán). The author thanks Prof. Tom Sundius for his permission to use MOLVIB.

References

1. S.A. Brandán, J. Mol. Struc. (THEOCHEM) **908**, 19 (2009)
2. S.A. Brandán, in *Structural and Vibrational Properties of Chromyl Perchlorate*, ed. by L.E. Mattews Chapter 3 (Nova Science Publisher, Inc., New York, 2010)
3. S.A. Brandán, M.L. Roldán, C. Socolsky, A. Ben Altabef, Spectrochim. Acta. Part A **69**, 1027 (2008)
4. A. Ben Altabef, S.A. Brandán, J. Mol. Struct. **981**, 146 (2010)

5. C.J. Marsden, K. Hedberg, M.M. Ludwig, G.L. Gard, Molecular structure of $CrO_2(NO_3)_2$ in gas phase: a novel form of coordination. Inorg. Chem. **30**, 4761–4766 (1991)

6. M.L. Campbell, S.B. Larson, N.K. Dalley, Acta Cryst. **B37**, 1741–1744 (1981)

7. P.H. van Rooyen, J.C.A. Boeyens, Acta Cryst. **B31**, 2933–2934 (1975)

8. M. Dobler, J.D. Dunitz, P. Seiler, Acta Cryst. **B30**, 2741–2743 (1974)

9. W. Xu, J.-M. Shi, X. Zhang, Acta Cryst. **E61**, m2194–m2195 (2005)

10. J.W. Bats, P. Coppens, Å. Kvick, Acta Cryst. **B33**, 1534–1542 (1977)

11. T.-H. Lu, J.-M. Lo, B.-H. Chen, M.-Y. Yeh, S.-F. Tung, W.-T. Huang, H.-.H. Yao, Acta Cryst. **C54**, 1067–1068 (1998)

12. W. Yong, J.-M. Dou, D.-Z. Zhu, Y. Liu, X. Li, P.-J. Zheng, Acta Cryst. **E57**, m127–m129 (2001)

13. V. Adovasio, M. Nardelli, Acta Cryst. **C51**, 380–382 (1995)

14. A.E. Reed, L.A. Curtis, F. Weinhold, Chem. Rev. **88**(6), 899 (1988)

15. J.P. Foster, F. Weinhold, J. Am. Chem. Soc. **102**, 7211 (1980)

16. A.E. Reed, F. Weinhold, J. Chem. Phys. **83**, 1736 (1985)

17. R.F.W. Bader, *Atoms in Molecules: A Quantum Theory* (Oxford University Press, Oxford, 1990). ISBN 0198558651

18. A.B. Nielsen, A.J. Holder, *GaussView User's Reference,* (GAUSSIAN, Inc., Pittsburgh, 2000–2003)

19. Gaussian 03, Revision B.01, M.J. Frisch, G.W. Trucks, H.B. Schlegel, G. E. Scuseria, M. A. Robb, J. R. Cheeseman, J. A. Montgomery, Jr., T. Vreven, K. N. Kudin, J. C. Burant, J. M. Millam, S. S. Iyengar, J. Tomasi, V. Barone, B. Mennucci, M. Cossi, G. Scalmani, N. Rega, G. A. Petersson, H. Nakatsuji, M. Hada, M. Ehara, K. Toyota, R. Fukuda, J. Hasegawa, M. Ishida, T. Nakajima, Y. Honda, O. Kitao, H. Nakai, M. Klene, X. Li, J. E. Knox, H. P. Hratchian, J. B. Cross, C. Adamo, J. Jaramillo, R. Gomperts, R. E. Stratmann, O. Yazyev, A. J. Austin, R. Cammi, C. Pomelli, J. W. Ochterski, P. Y. Ayala, K. Morokuma, G. A. Voth, P. Salvador, J. J. Dannenberg, V. G. Zakrzewski, S. Dapprich, A. D. Daniels, M. C. Strain, O. Farkas, D. K. Malick, A. D. Rabuck, K. Raghavachari, J. B. Foresman, J. V. Ortiz, Q. Cui, A. G. Baboul, S. Clifford, J. Cioslowski, B. B. Stefanov, G. Liu, A. Liashenko, P. Piskorz, I. Komaromi, R. L. Martin, D. J. Fox, T. Keith, M. A. Al-Laham, C. Y. Peng, A. Nanayakkara, M. Challacombe, P. M. W. Gill, B. Johnson, W. Chen, M. W. Wong, C. Gonzalez, J.A. Pople, *Gaussian 03, Revision B.04* (Gaussian, Inc., Pittsburgh, 2003)

20. A.D. Becke, J. Chem. Phys. **98**, 5648 (1993)

21. C. Lee, W. Yang, R.G. Parr, Phys. Rev. B **37**, 785 (1988)

22. G. Fogarasi, P. Pulay, in *Vibrational Spectra and Structure*, vol. 14, ed. by J.E. Durig (Elsevier, Amsterdam, 1985), p. 125

23. P. Pulay, G. Fogarasi, F. Pang, J.E. Boggs, J. Am. Chem. Soc. **101**(10), 2550 (1979)

24. T. Sundius, J. Mol. Struct. **218**, 321 (1990)

25. T. Sundius, in *MOLVIB: A Program for Harmonic Force Field Calculation, QCPE Program No. 604* (Adenine Press, Schenectady, 1991)

26. E.D. Glendening, A. E. Reed, J.E. Carpenter, F. Weinhold, NBO Version 3.1 (Theoretical Chemistry Institute, University of Wisconsin, Madison, 1996)

27. F. Biegler-König, J. Schönbohm, D. Bayles, AIM2000; a program to analyze and visualize atoms in molecules. J. Comput. Chem. **22**, 545 (2001)

28. C.J. Marsden, L. Hedberg, K. Hedberg, Inorg. Chem. **21**, 1113 (1982)

29. R.J. French, L. Hedberg, K. Hedberg, G.L. Gard, B.M. Johnson, Inorg. Chem. **22**, 892 (1982)

30. R.J. Gillespie (ed.), *Molecular Geometry* (Van Nostrand-Reinhold, London, 1972)

31. R.J. Gillespie, I. Bytheway, T.H. Tang, R.F.W. Bader, Inorg. Chem. **35**, 3954 (1996)

32. L. E. Sutton (ed.), *Tables of Interatomic Distances and Configurations in Molecules and Anions* (Burlington House, London, 1958), p. 49

33. K. Nakanishi, P.H. Solomon, *Infrared Absorption Spectroscopy* (Holden-Day, Inc., Sydney, 1977)